An Introduction to Mission Design for Geostationary Satellites

SPACE TECHNOLOGY LIBRARY

Editorial Board

An Introduction to Mission Design for Geostationary Satellites

by

J. J. Pocha

British Aerospace, Space and Communications Division,
Stevenage, U.K.

D. Reidel Publishing Company

A MEMBER OF THE KLUWER ACADEMIC PUBLISHERS GROUP

Dordrecht / Boston / Lancaster / Tokyo

Library of Congress Cataloging in Publication Data

CIP

Pocha, J. J. (Jehangir. J.), 1945–
An introduction to mission design for geostationary satellites.

(Space technology library)
Includes bibliographies and index.
1. Geostationary satellites. 2. Artificial satellites in telecommunication. I. Title II. Series.
TK5104.P63 1987 621.38′0423 87–4909
ISBN 90–277–2479–2

Published by D. Reidel Publishing Company,
P.O. Box 17, 3300 AA Dordrecht, Holland.

Sold and distributed in the U.S.A. and Canada
by Kluwer Academic Publishers,
101 Philip Drive, Assinippi Park, Norwell, MA 02061, U.S.A.

In all other countries, sold and distributed
by Kluwer Academic Publishers Group,
P.O. Box 322, 3300 AH Dordrecht, Holland.

Printed in The Netherlands

This work is dedicated to my wife, Meher,
and to my children, Cyrus and Shirin.

Foreword

James R. Wertz
Managing Editor, Space Technology Library

In 1945 Arthur C. Clarke first proposed the use of satellites in geostationary orbits as a means of achieving advanced communications. That concept has proved remarkably successful and now the geostationary ring is the single most common orbit for unmanned spacecraft. It is the cornerstone of a major communications revolution which has seen the cost of a telephone call between Europe and North America drop by a factor of more than ten in a fifteen year period. Geostationary satellites have brought modern communications to many areas of the world where they had never been before, and brought the world closer together for all of us. Nearly instantaneous telecommunications from around the world have become so commonplace as to go practically unnoticed on the evening news.

The effective utilisation of this major resource requires an understanding of both satellite technology and the mission planning, analysis, and operations required to launch and maintain geostationary satellites. It is this latter area which is clearly presented to us by Jehangir Pocha in this volume. Dr. Pocha provides us with the basic information needed to analyse, plan, and carry out a geosynchronous mission from launch, through orbit transfer and station acquisition, to station-keeping and on-orbit operations. Dr. Pocha's many years with the Dynamics Department of British Aerospace's Space and Communications Division, a major supplier of geostationary satellites for the world market, have provided him with the practical experience so essential to understanding the issues involved. This book serves as a valuable guide to anyone interested in geostationary satellites, from those approaching space technology for the first time to experienced system engineers in need of a ready reference for this important class of missions.

This book inaugurates a new series by D. Reidel – the *Space Technology Library*. In most areas of science and technology, leadership and knowledge have come from the academic community where the preparation of technical books is a well-established tradition. Largely because of the high costs of space exploration, much of the progress in space technology over the last 30 years has come from industry, and government laboratories. Space technology does not have the wealth of strong technical books which provide the intellectual foundation for most other fields of science and engineering. It is our hope that the *Space Technology Library*, in conjunction with individuals and groups from far-sighted organisations such as BAe, can begin to fill this major gap in the technical literature. Providing as it does the analytic basis for the most fully developed and utilised type of space mission, this volume is a superb and most appropriate beginning.

Table of Contents

Preface

This text first saw life as a training document for mission analysts within the Dynamics Department of British Aerospace's Space and Communications Division. The intention was to make available in an easily understandable form, to newcomers to the field, the techniques and experience accumulated through years of mission design for a wide variety of spacecraft. It soon proved its worth, and not only with prospective mission analysts. Systems engineers, and others generally interested in geostationary spacecraft found it to be of value. It was clear that the text satisfied a need that could not otherwise be filled. There are many, and very good, books on celestial mechanics, there are many excellent books on astrodynamics, but until now there has been no book that explains the ins and outs of mission design, explains *why* things are done the way they are, as well as *how* they are done.

For the much wider audience for which this book is designed, much of the material has been extensively revised, and portions have been completely re-written with the objective of providing a detailed and well-rounded account of the mission design of geosynchronous spacecraft. The work will be of interest not only to practising mission analysts, but also to spacecraft systems engineers, spacecraft project managers and to those who wish to employ the unique attributes of geosynchronous spacecraft for useful purposes.

A basic understanding of orbit mechanics is essential for a good understanding of the text, but even this may be dispensed with for an appreciation of the introductory sections of the various chapters. This book is complementary to works on celestial mechanics, astrodynamics etc, and not a substitute for them. Fundamental orbit mechanics concepts are, therefore, taken for granted, and only those developments are presented which have a particular application to geostationary spacecraft.

For reasons of logical development and narrative flow, the various mission design activities are presented in the order in which the mission of a real spacecraft will unfold. This does not imply that the analyses are, or should be, carried out in this order in support of a spacecraft project. No such simple ordering is possible in a field as complex as this, and in any case, the emphasis and ordering of mission design activities are governed by the requirements of a particular spacecraft project.

After an introductory chapter, setting the scene and presenting a profile of the mission of a geostationary spacecraft, Chapters 2 to 6 take the spacecraft from launch through to the end of its useful life in geostationary orbit. The remaining chapters address topics that arise in several mission phases without being specific to any one.

Throughout the text no distinction is made between the meaning of the words *spacecraft* and *satellite*, and they are used interchangeably, and the same applies to the words *geostationary* and *geosynchronous*. By deliberate policy, the text is left free of references. The final section of each chapter contains an annotated list of the most useful literature pertaining to the subject of that chapter.

The author gratefully acknowledges British Aerospace for valuable assistance in the preparation of the camera-ready copy, and the European Space Agency (ESA) for permission to cite the ESA family of communications spacecraft as examples throughout the text. On a more personal note, the author is indebted to his wife, Meher, for proof-reading every page of the text and enhancing its readability, to Jacqui Dennis for the typing and page layout, and to Jim Wertz, the Chief Editor of this series, for his unflagging enthusiasm for this book, and his valuable suggestions for its improvement. If, in spite of all this help, errors remain, they are entirely the author's own responsibility.

Stevenage, Hertfordshire
December, 1986

Jehangir J. Pocha

List of Acronyms

Acronyms are commonly used in aerospace literature. The following list explains the ones used in this text.

ABM	Apogee boost motor (solid propellant)
AEF	Apogee engine firing (relating to liquid propellant apogee engines)
AMF	Apogee motor firing (relating to solid propellant apogee motors)
AND	Active nutation damping
BOL	Beginning of life (mass of the spacecraft)
CNES	Centre Nationale d'Etudes Spatiales (French national space agency)
COMSAT	Communications Satellite Organisation (U.S.)
ECS	European Communications Satellite (an ESA-sponsored satellite used for European communications traffic)
ESA	European Space Agency
GEO	Geostationary Earth orbit
GSFC	Goddard Space Flight Centre (of NASA)
GTO	Geosynchronous transfer orbit
INMARSAT	International Maritime Satellite Organisation
INTELSAT	International Telecommunications Satellite Organisation
IUS	Inertial Upper Stage (for *Shuttle*)
MARECS	Maritime ECS (an ESA-sponsored satellite used by INMARSAT)
NASA	National Aeronautics and Space Agency (of the U.S.)
OTS	Orbital Test Satellite (an ESA-sponsored precursor to the ECS/MARECS family of satellites)
PAM-D/D2	Payload Assist Module (*Shuttle* upper stages)
PO	Parking Orbit (of *Shuttle*)
RAE	Royal Aircraft Establishment (of the U.K.)
RCS	Reaction control system (on spacecraft)

SGLS	Space–Ground Link Subsystem (of the U.S. Air Force)
SHF	Super–high frequency
STDN	Space Tracking and Data Network (of NASA)
TTC	Tracking, Telemetry, and Command (subsystem of the satellite)
UHF	Ultra–high frequency
VHF	Very–high frequency

CHAPTER 1

INTRODUCTION

Geostationary communications spacecraft are now relatively commonplace, and a significant and rapidly growing fraction of the world's telecommunications traffic is routed through them. The principal characteristic of the geostationary orbit (GEO) that makes it so suitable for communications spacecraft, is that the spacecraft position is essentially fixed relative to an Earth-fixed observer. This has the great advantage that continuous coverage of a certain portion of the globe is maintained throughout the satellite's lifetime.

For an orbit to be geostationary it must be circular, it must lie in the Equatorial plane, and it must have a period exactly equal to the Earth's rotation period. The last requirement results in an orbit radius of approximately 6.61 Earth radii. The orbit is, therefore, a high one, much higher than the orbits of most Earth-orbiting scientific satellites, and a great deal of energy is required to place a satellite into such an orbit. The orbit geometry is shown in Figure 1.1. In this section we shall outline very briefly the mission characteristics of communications spacecraft in order to set in perspective the more detailed treatments of each aspect, which follow in subsequent sections. Figure 1.2 shows a diagrammatic representation of a typical geosynchronous mission.

Placement of a satellite into a geostationary orbit is usually achieved by a combination of a multi-stage launch vehicle, and perigee and apogee motors. The launch strategy depends upon the launch vehicle, of which there are several in use around the world. Most are of the multi-stage expendable type, which place the spacecraft into an inclined, highly elliptic intermediate orbit called the geostationary transfer orbit (GTO). The inclination of the transfer orbit is usually determined by the latitude of the launch site, and the ellipticity is so designed as to place the apogee of the orbit at synchronous radius and very near the Equatorial plane. The perigee height of the transfer orbit is determined by the requirement to avoid excessive perturbation due to the effects of aerodynamic drag. At the apogee of the transfer orbit, the apogee motor is fired in such a direction as to achieve an Equatorial synchronous orbit.

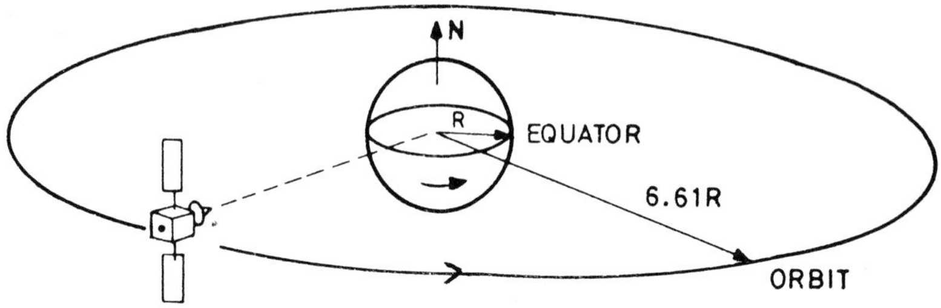

- EARTH'S ROTATION PERIOD = ORBIT PERIOD
- SPACECRAFT STATIONARY WITH RESPECT TO EVERY POINT ON THE EARTH.
- ORBIT LIES IN THE EQUATORIAL PLANE

Fig.1.1 Characteristics of the Geostationary Orbit

An exception to the above strategy is the US Space Shuttle. This is a manned and reusable vehicle which carries a large payload into a range of near-Earth orbits. To launch a geosynchronous spacecraft, the Shuttle carries the spacecraft and a boost stage into a suitable close-Earth orbit. The satellite and boost stage combination are then deployed from the Shuttle's cargo bay. At an appropriate time the boost stage is fired to set up a transfer orbit similar to that described above. At the apogee of the transfer orbit, the satellite's apogee boost motor is fired to achieve geosynchronous orbit.

Before a satellite can be launched, however, the allowed time for launch – the *launch window* – must be determined. The launch window is usually constrained by the requirements imposed by the satellite's systems, primarily the power and thermal control systems. If the satellite is spin-stabilised during the transfer orbit, the thermal control system will impose limits on the Sun direction with respect to the satellite spin axis, to prevent continuous strong illumination of parts of the satellite. To allow adequate power generation during transfer orbit, the Sun direction must be such as to ensure strong illumination of the solar array panels, and the eclipse periods must not be too long. To allow adequate attitude determination, the Sun/satellite/Earth geometry must satisfy certain requirements. The simultaneous satisfaction of all such constraints results in the spacecraft being allowed to be launched only within specific time slots. In order to allow the launch campaign to be reasonably flexible, one tries to provide at least one launch opportunity for every day of the year, but sometimes this is not possible, e.g. the launch window of the Maritime Communications Satellite, MARECS, is closed for a 40 day season during the year.

Once the launch has successfully taken place, the events during the transfer orbit are carefully organised with a prime aim in view – to achieve as accurate an apogee manoeuvre as possible. For the apogee manoeuvre to be executed correctly, the satellite must be manoeuvred into the desired attitude. This can be done by reaction control thrusters mounted on the satellite. For attitude stability during the apogee burn, the spacecraft must be passively stable or actively stabilised. To achieve the former, the spacecraft is spun about the nominal apogee boost motor thrust vector axis. This ensures that any perturbations due to thrust misalignment during motor firing are sufficiently small to ensure that the magnitude of the total impulse in the required direction is adequate.

Once the apogee manoeuvre has been performed satisfactorily, the satellite will be in a near-synchronous orbit, also known as the drift orbit. The task still remains, however, to bring it to the correct on-station operational longitude. This will, in general, be different from the longitude at which the motor is fired, though not far from it. The manoeuvres required to bring the satellite on to station, constitute the *station-acquisition* strategy, the objective of which is to achieve station within a specified time limit, in the most efficient manner. The manoeuvres are implemented by the on-board reaction control system, for which sufficient fuel must be available. Owing to the errors inherent in the launch procedures, the orbit characteristics will be non-nominal.

The station acquisition manoeuvres, therefore, depend upon the particular set of drift orbit characteristics that pertain after the apogee manoeuvre. In order to plan the fuel allowance for station acquisition, it is necessary to proceed on a statistical basis, and allow, for example, enough fuel to make it possible to acquire station with a probability of 99%. To determine the fuel requirement, a Monte Carlo procedure is used.

When the satellite, after the execution of station acquisition manoeuvres, arrives at the required longitude, it must be maintained there within specified limits for the duration of its operational lifetime. The manoeuvres required to achieve this constitute the *station-keeping* strategy. Owing to the effects of the Earth's gravitational potential, solar radiation and the gravitational attraction of the Sun and the Moon, a satellite initially in a perfectly geostationary orbit will tend to lose synchronism due to the perturbation of the orbit elements, e.g. semi-major axis, eccentricity, and inclination. This will result in the satellite moving relative to an Earth-fixed observer. If this movement is to be kept within strict limits, as it is usually required to be, station-keeping manoeuvres must be carried out through the lifetime of the satellite.

Throughout the satellite's lifetime, the execution and planning of manoeuvres requires a knowledge of the spacecraft's orbit and attitude. Orbit determination during the mission is the responsibility of the ground station network, but the accuracy with which the orbit is to be determined is set by mission considerations. Once the tracking accuracy requirements are set, the tracking campaign necessary to satisfy these requirements is determined through a covariance analysis embracing nominal orbit characteristics, the nature and frequency of the tracking observations, and the statistical characteristics of the errors inherent in each observation. The tracking campaign is itself constrained by the coverage of the satellite available from a given set of ground stations. The results of the covariance analysis will indicate if the requirements can be satisfied or not. The requirements of the tracking campaign can influence the event sequence of the mission, specially strongly in the transfer orbit, when ground station coverage is intermittent.

In the discussion above, we have, very briefly, described the events that occur in the lifetime of a communications spacecraft. A detailed analysis of these events constitutes the mission design activity for a spacecraft. It is the objective of such analyses to:

- assess the compatibility and integrity of the mission as a whole;
- develop an event chronology for the complete mission;
- define the fuel budget to execute the mission with a specified probability of success;
- define aspects which lead to satellite hardware decisions;
- determine and define the requirements relating to the control of the satellite from the ground.

The mission analysis task begins at the start of system conception, proceeds through the study and design phases, and carries right through into post-launch mission support. The investigations of various aspects of the mission are generally brought together in a document called the *Mission Manual*, which is periodically updated, and which forms the reference document for all mission matters. During the mission analysis work, close liaison is maintained with satellite system and subsystem designers to identify the mutual interactions between these and the mission, at the earliest possible stage.

It will be seen in later sections that mission design can be computationally quite complex. The aid of the digital computer is, therefore, extensively invoked. Computer programs are required for tasks such as orbit optimisation, launch window derivation, station acquisition, station-keeping, orbit propagation, ground station coverage, etc. Since the analysis of one mission phase naturally leads into the next, it will be found useful to design the programs to work as a suite, with the output of one program being so designed as to provide directly the input of another. Not only is this more efficient in terms of time and effort, but it also minimises data error.

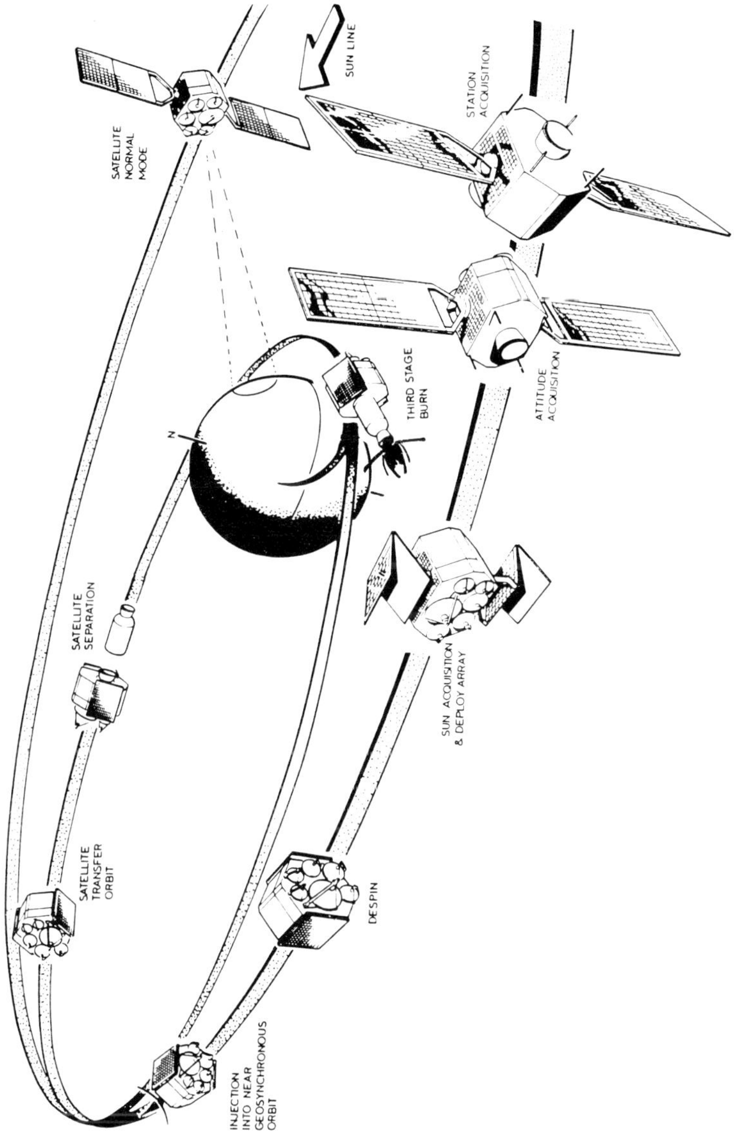

Fig.1.2 Schematic of the Complete Mission

1.1 Further Reading

Reference (a) provides a good introduction to geostationary communications spacecraft. Though the main emphasis is on the communications payload, the mission aspects are also briefly covered.

Reference (b) provides a glimpse into the future development of communications spacecraft.

Reference (c) provides a general introduction to the orbit mechanics and operational aspects of geostationary orbits, and is a very useful supplement to this text.

(a) Martin, J. : 1978, "Communication Satellite Systems", published by Prentice Hall.

(b) Sachdev, D.K. : 1981, "Satellite Communication Technology – Challenges for the 1980s", *J of Spacecraft and Rockets*, Vol.18, No.2, pp.110–117.

(c) Soop, M. : 1983, "Introduction to Geostationary Orbits", *ESA SP-1053*.

CHAPTER 2

LAUNCH

In this section we shall consider the mission aspects of placing a satellite into transfer orbit. Two main activities arise, that of selecting the launch vehicle appropriate to the satellite and the mission, and that of assessing the launch window. In fact the first activity usually occurs the other way around! Because only a few launch vehicles are available for selection, a choice is made very early in the project mainly on grounds such as lift capability, cost, dual compatibility etc. The satellite is then designed to the constraints imposed by the launch vehicle, and its associated launch site.

Section 2.1 contains a discussion on the capabilities and characteristics of launch vehicles, while Section 2.2 covers the assessment of the launch window.

2.1 Launch Vehicles

2.1.1 Introduction

The great and increasing popularity of the geosynchronous earth orbit (GEO) has led to a number of launch vehicles being developed to satisfy the needs of GEO missions. The very proper desire on the part of larger nations to be self-sufficient in the means of exploiting the synchronous orbit has led to large development programmes resulting in, or planned to result in, indigenous launch capability. The needs of Western European countries has been satisfied by the *Ariane* family of vehicles, the product of French foresight and determination, and Europe-wide collaboration under the aegis of the European Space Agency, (ESA). In the U.S. the *Space Transportation System* – "the*Shuttle*" – provides the very great proportion of services to geostationary spacecraft, though the older, more conventional launch vehicles remain important and plans are afoot to market their services to spacecraft users on a commercial basis. The growing space aspirations of Japan are reflected in its large space research and development programme. Already the *N2* vehicle,

Table 2.1

Launch Vehicle Characteristics

Launch vehicle / Upper Stage	Country of Origin	Stabilisation Mode	Payload (kg)	Orbit	Comments
Delta 3920/PAM-D	U.S.	I for stages 1 & 2 SS for stage 3	1280	GTO; i = 27°	stages 1 & 2 liquid prop. stage 3 solid prop.
Atlas-G/Centaur	U.S.	I for all stages	2360	GTO; i = 25°	stages ½ and 1 liquid prop. stage 2 cryogenic
Shuttle/PAM-D	U.S.	SS upper stage	1250	GTO; i = 27.3°	solid prop. upper stage
Shuttle/PAM-D2	U.S.	SS upper stage	1860	GTO; i = 27.3°	solid prop. upper stage
Shuttle/IUS	U.S.	I upper stage	2270	GEO	solid prop. upper stage
Shuttle/TOS	U.S.	I upper stage	2730–5910	GTO	solid prop. upper stage
Ariane 3	Europe	I for all stages	2580	GTO; i = 7°	liquid prop. for stages 1 & 2
Ariane 4 family	Europe	I for all stages	1900–4200	GTO; i = 7°	cryogenic prop. for stage 3
Ariane 5 (planned)	Europe	I for both stages	5200–8000	GTO	2 cryogenic stages plus solid boosters on 1st stage
N2	Japan	I for stages 1 & 2 SS for stage 3	355	GEO	stages 1 & 2 liquid prop. stage 3 solid prop.
H1	Japan	as for N2	550	GEO	as for N2 but 2nd stage is cryogenic (LOX/LH_2)
Long March (CZ-3)	China		1400	GTO; i = 31.1°	stages 1, 2 & 3 liquid prop.
Proton	U.S.S.R.	I for all stages	2200	GEO	

Notes: I = Inertial or 3 axis stabilised; SS = Spin stabilised; i = Orbit inclination; prop. = propellant; GTO = geosynchronous transfer orbit; GEO = geosynchronous orbit

stemming from U.S./Japanese collaboration, has been used successfully to orbit a series of experimental communications satellites. A more capable vehicle, the *H1* is under development. The avowed intention of India to be self–sufficient in Space has led to a substantial launcher development and test programme, which, from small beginnings, is intended to result in a substantial capability into GEO. China has already used indigenous hardware – the *Long March* vehicles – to launch its own experimental GEO communications satellites.

It is clear, therefore, that in the very near future several launch systems will be competing for the lucrative business of launching geosynchronous spacecraft. As the number of geosynchronous spacecraft required to meet the varied and growing needs is expected to increase very rapidly, there should be ample opportunities for most of the launch systems, provided costs can be kept in control. Table 2.1 lists most of the available vehicles, the type of orbit into which the spacecraft is placed, and, where available, the payload capability of the vehicle. In the section below we shall consider some of the mission–related characteristics of the more frequently used vehicles. Unfortunately, although the Russian *Proton* falls within this category, lack of detailed information upon it prevents a useful discussion. For further details of launch vehicles their User Manuals should be consulted. These usually give brief details of the launch vehicle hardware, and full information on performance capability, mechanical and electrical interfaces, safety requirements, documentation, launch campaign, and launch site facilities.

Selection of the launch vehicles places a number of constraints upon the physical and mission characteristics of the satellite. The physical characteristics are governed by the vehicle's payload capability, the volume available within its payload compartment, the environment imposed upon the satellite, the strength limitations of the vehicle structure, and the mechanical and electrical interface requirements between the vehicle and the satellite. The mission characteristics are imposed by the vehicle's trajectory, its accuracy and the stability requirements of the vehicle and the satellite.

2.1.2 Launch Vehicle Characteristics

Launch vehicles, with the exception of the *Shuttle*, are multi–stage, expendable vehicles which are capable of imparting to their payloads an increment of velocity which depends essentially upon the mass of the payload. Different launch vehicles have different launch profiles. The *Delta* vehicles use three stages for the achievement of the transfer orbit. The second stage is restartable, and an extended coast period occurs during second stage flight. At the end of the first burn of the second stage, the vehicle is in a 185 km altitude circular orbit. The subsequent coast phase takes the vehicle to a position very close to the Equatorial plane. At this point, the second stage is re–ignited, and burns out. Stage separation then occurs, and the unguided, spin stabilised third stage places the satellite into the required transfer orbit whose perigee is nominally over the Equator. The implication of such a strategy is that the inclination of the transfer orbit can be modified slightly ($\sim 2^{o}$) from that of the circular orbit, without affecting the position of the transfer orbit perigee. This flexibility is useful when an inclination change is required, to optimise the transfer orbit. The launch trajectory is shown in Figure 2.1.

Ariane, on the other hand, has a slightly different launch profile. It is also an expendable three-stage vehicle but has no coast phase. The trajectory is a fully guided one (as opposed to guidance for just the first two stages of the *Delta* vehicle), and if the transfer orbit inclination is altered, the position of the transfer orbit perigee changes too. This slightly complicates optimisation of the transfer orbit. The *Ariane* launcher, however, has significant advantages over the *Delta* vehicle. In the first place, launch occurs from Kourou (French Guyana) rather than from Cape Canaveral (USA). Since Kourou is at a much smaller latitude, the inclination of the transfer orbit is 7° instead of 28.5^{o}. The plane change manoeuvre performed by the satellite at the apogee of the transfer orbit is, therefore, much smaller. This results in a larger useful mass on station, all other things being equal. Secondly, as the trajectory of *Ariane* is fully guided, transfer orbit dispersions are much smaller than is possible with the unguided *Delta* third stage. This results in a smaller fuel allocation for the station acquisition manoeuvres, and hence in greater useful mass on station. The launch trajectory for *Ariane* is shown diagrammatically in Figure 2.2

The *Space Shuttle* has a launch profile that is significantly different from that of the two expendable launchers discussed above. It is a manned re-usable vehicle with a very much greater payload capability. The manned stage is, however, limited to near-Earth orbits. To place a communications satellite into transfer orbit, a perigee boost stage is required. It is the function of the perigee stage to place the satellite into transfer orbit or geosynchronous orbit. Several upper stages are available and more are currently under development. For small spacecraft, i.e. in the 1200 kg to 1800 kg class, the McDonnell Douglas *PAM-D/D2* stages are available. These are simple, unguided, spin-stabilised, solid propellant stages which place the spacecraft into GTO. For larger payloads Boeing's *Inertial Upper Stage* is available. This is a fully-guided two-stage vehicle and places the spacecraft directly into GEO. Other stages under development include the Orbital Sciences Corporation *Transfer Orbit Stage*, and those designed specifically for the Intelsat 6 and Hughes Leasesat spacecraft. All these are solid propellant stages which place the spacecraft into GTO. Liquid propellant upper stages have also been designed by TRW and McDonnell Douglas but have not at the time of writing, entered the production stage. The parking and transfer orbits that arise with a *Shuttle* launch are shown diagrammatically in Figure 2.3.

Another difference between the launch characteristics of various vehicles is the attitude in which the spacecraft is left at burn-out of the final stage, and the spin rate imparted to the satellite by the launcher. The *Delta* vehicle having a spinning but unguided third stage leaves the satellite spinning at around 60 rpm, with the spin axis attitude being that of the third stage at burn-out. The *Ariane* vehicle has the capability of leaving the satellite at any specified attitude after injection into transfer orbit. It can also spin the payload up to a maximum of 5 rpm using existing equipment (higher spin rates are possible using user-supplied spin tables which are payload-chargeable). The situation with the *Shuttle* is more varied and depends upon the perigee stage being used and upon the airborne support equipment. These considerations have an important impact upon transfer orbit manoeuvres.

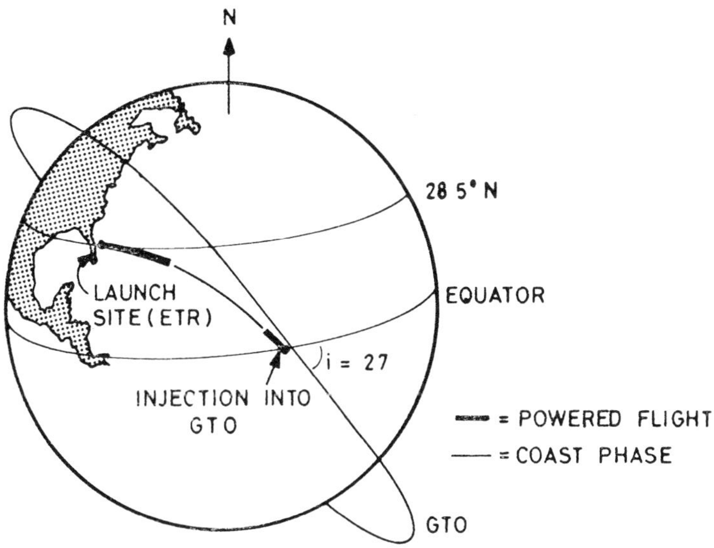

Fig.2.1 Delta Launch Trajectory

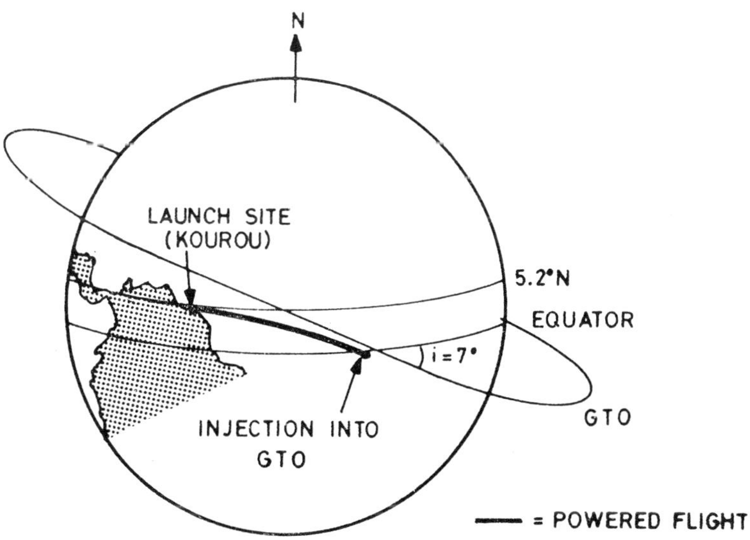

Fig.2.2 Ariane Launch Trajectory

For example, with a *Delta* launch, the spin axis re-orientation, or slew, manoeuvre is of the order of 130°, while with *Ariane*, this is reduced to a trimming manoeuvre of about 6°. On the other hand, the low spin rate imparted by *Ariane* requires the inclusion of a spin-up manoeuvre to satisfy apogee motor firing requirements. For launch by the *Shuttle/PAM-D* combination, the transfer orbit manoeuvres are similar to those appropriate to a *Delta* vehicle.

Shuttle launches, however, require us to consider an extra mission phase, the parking orbit phase. This is the phase between the separation of the spacecraft-plus-upper stage stack from the Orbiter, and the burn-out of the upper stage motor. Taking the *PAM-D* upper stage as an example, the event sequence for this phase is shown in Table 2.2. Orbiter-related safety considerations impose certain requirements and restrictions upon the manoeuvres that can be carried out. For example, attitude control thruster actuation is not allowed before a certain safe separation distance is achieved between the Orbiter and the spacecraft. The safe separation distance varies with the size of the thruster and the nature of the propulsion system employed. Similarly, the upper state is only allowed to commence firing after a much greater separation distance is achieved – for the *PAM-D* this implies a 45 minute coast phase between separation from *Shuttle* and ignition of the upper stage motor.

A most important activity associated with the parking orbit is the attitude control of the spacecraft-plus-upper stage stack. The deployment and configuration characteristics of the *PAM-D/D2* result in the spacecraft/upper stage combination being a spinning, prolate body. Such a body is basically unstable and must be stabilised artificially. The control system to do this is called an *Active Nutation Damper*, (AND), descriptive of its main function. Active nutation damping is achieved through sensing the body rates, and feeding energy into the system through appropriate actuators (thrusters, reaction wheels, etc.) obeying carefully designed control laws. It is the function of the AND system to hold the spin axis attitude to sufficient accuracy for an efficient upper stage burn to be assured. The AND can be implemented either by using equipments and systems already on board the spacecraft, or by a purpose-built system mounted on the upper stage. The former is generally adopted, as it is more mass-efficient.

Design of the AND is further complicated, for recent spacecraft, by the presence of large quantities of liquid propellants on board. Use of efficient bi-propellant propulsion systems for the apogee manoeuvre and the attitude and orbit control functions, results in over half the total spacecraft mass being liquid propellants. The presence of finite ullage in the propellant tanks allows bulk motion of the fluid and viscous dissipation of energy at the tank surface boundary layer. These mechanisms interact with the rigid body motion to produce a complex motion, which needs to be well understood before a control system can be designed to hold the spin axis attitude within desired limits. The mission design aspects of spacecraft stability are discussed in Chapter 10.

It can be seen that parking orbit mission design must take full account of *Shuttle* constraints, upper stage characteristics, and the control characteristics of the spacecraft-plus-upper stage stack.

Table 2.2

Shuttle/PAM-D Parking Orbit Sequence

Time (min : sec)	Event
T - 60:00	Turn on Shuttle PAM-D power
T - 40:00	Start manoeuvre to deployment attitude
T - 28:30	Achieve deployment attitude
T - 20:00	Start mechanical sequence
T - 19:28	Sunshield opening complete
T - 17:28	Restraints withdrawal complete
T - 17:18	Spin-up complete
T - 17:00	Spacecraft Airborne Support Equipment ON
T - 15:00	Spacecraft configured
T - 10:00	Spacecraft to internal power
T - 3:00	Start terminal sequence
T - 00:00	Deploy PAM – sequencing system initiated by separation switches
T + 2:05	Enable Active Nutation Damping (AND) system
T + 44:56	Arm ordnance bus
T + 44:58	Disable AND system
T + 45:00	Solid motor ignition
T + 46:25	Solid motor burnout
T + 48:20	Spacecraft separation
T + 48:23	PAM tumble system activation

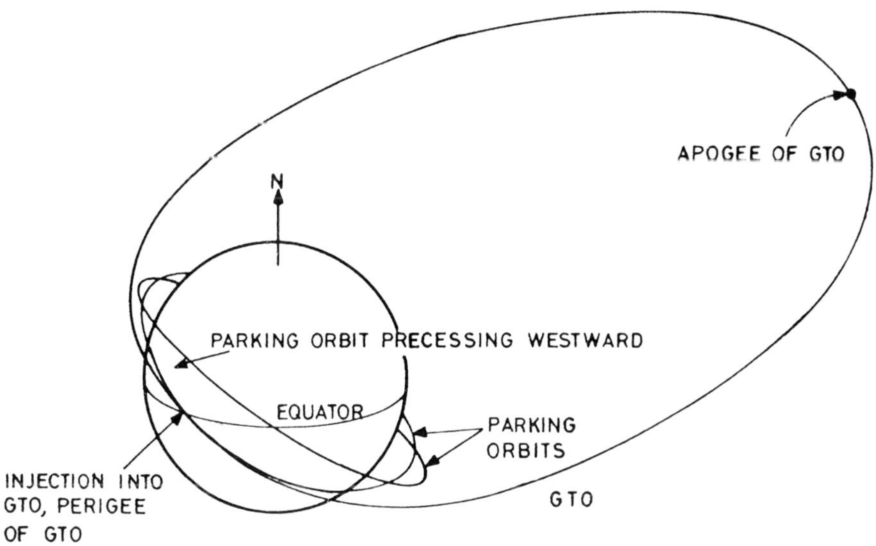

Fig.2.3 Parking and Transfer Orbits for Shuttle

2.1.3 Reliability

All space vehicles are designed to certain reliability goals. Unlike satellites, the same type of which only a few are built, the design reliability goals of launch vehicles can be verified in practice owing to the large statistical sample available for examination. For the satellite designer, the reliability figure of interest is the reliability with which a launch vehicle can place a payload into GTO or GEO. For the *Delta* vehicle the reliability over its lifetime is 93.3%. During the period of the last eight years the reliability is almost 98%, indicating the technical maturity of the design. The reliability design goal of the *Ariane* vehicle is 95%, though this has yet to be achieved, owing largely to the relative newness of the design. The reliability of the *Proton* vehicle is almost 93% over its lifetime. During the period of the last eight years the reliability has risen to almost 97%, again indicating the maturity of the design. The *Space Shuttle*, being a manned vehicle, has a design reliability of nearly 100%. However, the reliability of launch to transfer orbit must include the reliability of the perigee stage. The overall reliability is therefore the product of the reliabilities of the *Shuttle* and the perigee stage.

A significant new factor that has been introduced by the *Shuttle* is the possibility of satellite in-orbit check-out and repair. Once the Orbiter has reached parking orbit, it will open its cargo bay doors and commence deployment of the satellite/perigee stage combination. By use of the remote manipulator system, the Orbiter crew will be able to orient the satellite to the attitude required for perigee motor firing. The satellite can then be checked out for the correct functioning of its systems. If the satellite status is okay, the mission can proceed. If a fault is detected, it will be possible to abandon the mission, retrieve the satellite/perigee stage combination, stow it back in the cargo bay, and return it to Earth for investigation and fault correction. A limited satellite repair capability has already been demonstrated on orbit. The retrieval of damaged satellites, and their return to Earth has also been demonstrated.

In many cases it will be a requirement to design the spacecraft to be compatible with both the *Shuttle* and a conventional launch vehicle, e.g. *Ariane*. This will involve assessing two different mission profiles and designing the satellite to be suitable for both.

The tracking of, and data reception from, the vehicle during the launch phase is the responsibility of the launch authority, e.g. NASA, ESA, etc. and a contract to cover this is usually part of the launch negotiations. This aspect is, therefore, not a part of the mission analysis for the satellite, and will not be discussed here.

2.2 The Launch Window

2.2.1 Launch Window Constraints

The launch window of a satellite to be placed in geostationary orbit is determined by the simultaneous satisfaction of a number of constraints imposed by the satellite's systems. The most important constraints (important for their impact upon the launch window) are those on the solar aspect angle and upon eclipse durations. For the purpose of this discussion we shall consider the spacecraft to be spin-stabilised during the transfer orbit phase of the mission, and hence shall express various *aspect angles* with respect to the satellite's spin axis direction. For example, the solar aspect angle is the angle between the spacecraft's spin axis and the spacecraft-to-Sun direction. As the spin axis direction changes during the transfer orbit, due to re-orientation manoeuvres, the aspect angle constraints must be satisfied for all the attitudes that the spin axis will take up. The spin axis attitude must be changed from its injection attitude to its apogee motor firing (AMF) attitude through the execution of one or more slew manoeuvres. We have, therefore, to consider intermediate attitudes in addition to the injection and AMF attitudes. Below we shall outline the various constraints that arise.

The Solar Aspect Angle Constraint

The solar aspect angle must be constrained to lie within specified limits of the spin axis direction in order to avoid excessive continuous heating of parts of the satellite, and to ensure adequate power generation during transfer orbit. For example, for the OTS, MARECS and ECS spacecraft, the solar aspect angle was required to lie within $90^\circ \pm 25^\circ$ of the nominal spin axis direction, and this is typical of most communications satellites. This range is, however, further reduced to make allowances for tip-off error and attitude reconstitution error. The former arises from the imperfect (i.e. slightly asymmetric) separation of the satellite from the final stage of the launch vehicle, leaving the spin axis direction at a non-nominal value. The latter arises from the imperfect knowledge of the spacecraft attitude as determined on the ground. For the OTS spacecraft the combined effect of this is around $\pm 4^\circ$, resulting in a solar aspect angle constraint of $90^\circ \pm 21^\circ$ of the nominal spin axis direction.

The Eclipse Duration Constraint

The eclipse duration must not be too long, and must not occur during important orbit operations. During eclipses the satellite is powered only by its batteries, and as these are designed to deliver only a certain amount of energy, the maximum eclipse duration is limited. In addition, during eclipse, the satellite is shielded from the heating effect of the Sun and therefore cools down. To maintain desirable thermal conditions, the eclipse duration must be limited. The attitude reconstitution procedure requires knowledge of the Sun aspect angle. In eclipse this information is not available, and hence the position in the orbit during which eclipse occurs must not conflict with orbit positions at which attitude reconstitution is required. For example, for ECS, the eclipse duration must be less than 30 minutes per orbit, and eclipses must not occur between true anomalies of 70° and 190°.

The Attitude Determination Constraint

In order to reconstitute the satellite's attitude to the required accuracy, the Sun–Earth–satellite geometry must satisfy certain requirements. This translates to certain requirements upon the solar aspect angle. It so happens that for the SKYNET 4 family of spacecraft, the use of multiple methods of attitude reconstitution removes constraints upon the solar aspect angle. In general, however, this constraint must be borne in mind and applied when necessary. The attitude reconstitution process requires information about the Earth aspect angle, and this it receives from the output of the satellite's Earth sensor. This sensor will, however, be blinded when the Sun – a much more powerful radiation source than the Earth – approaches within a specified angle to the centre of its field of view. If, therefore, the Earth lies close to the Sun, as seen by the sensor, it will be unable to return Earth aspect angle information. The blinding effect persists for a short time before the sensor recovers its sensitivity; and as the satellite is spinning, this time lag translates to an angular distance. The angular separation between the Earth and the Sun as seen by the sensor is, therefore, required to be greater than a specified amount. This imposes a further constaint upon the Sun position and must be accounted for in the launch window analysis. For MARECS the requirement is that there should be no Sun–blinding of the Earth elevation sensor for two hours prior to AMF. For ECS the requirement is that there should be no Sun–blinding between true anomalies of 160° and 190°, which span the AMF position.

The Lunar Eclipse Constraint

A further constraint on the launch window is imposed by the Moon. The Moon has two main effects. It can eclipse the Sun from the satellite, and hence must be subject to similar constraints as for the eclipse duration discussed above. It can *confuse* the Earth sensor, and hence must be subject to similar constraints as for Sun–blinding. As the Earth sensor beam cannot discriminate effectively between the Earth and the Moon, misinterpretation of its output can result. For MARECS, the launch window is constrained by the requirement for there to be no lunar eclipses during transfer orbit. This constraint is not unduly restrictive on the launch window.

Launch Vehicle Constraints

An additional constraint on the launch window arises from the launch count-down sequence itself. The launch vehicle normally requires a minimum slot length of 20 minutes to allow for various holds during the count-down procedure. This was the value used by OTS, launched by *Delta 3914* from the Cape Canaveral complex in Florida. A value of 45 minutes has been adopted for *Ariane* launches from Kourou, French Guyana. In some cases it is a requirement that there be an open launch slot for a certain number of days after the selected launch date. This further shrinks the available launch window.

All the above constraints are super-imposed and simultaneously satisfied to derive the overall launch window for the satellite. If the launch takes place in conformity with the launch window, a successful launch will result in all the constraints being satisfied as required.

2.2.2 Launch Window Derivation

The calculation of the launch window depends upon the mission characteristics of the spacecraft. For OTS, for example, there was only one set of nominal transfer orbit parameters, and these were employed in the assessment. The step-by-step synthesis of the OTS launch window is shown in Figures 2.4 to 2.9.

For MARECS, a node-shifting capability was required to open the launch window adequately. This resulted in a set of three transfer orbits which had to be considered in the launch window analysis. The complete launch window is illustrated in Figure 2.10. Since the derivation of the launch window for a mission incorporating a node-shifting strategy is rather more complex than that for one without, we shall discuss it more fully here. Node-shifting is carried out in conjunction with a free-drift North-South station-keeping strategy (see Section 6.3.3.2), and is necessitated by the requirement to have a launch window open for most of the year. The node-shifting strategy is closely related to the choice of transfer orbit parameters, and is therefore covered in some depth in Section 3.2. Here we shall outline briefly the main points.

The reason that node-shifting opens the launch window is as follows. Without node-shifting, the ascending node of the post-AMF drift orbit will be the same as that of the transfer orbit, and because the launch trajectory is more or less fixed, the transfer orbit ascending node will occur at an essentially fixed longitude *with respect to the Greenwich meridian*. Now, the pole of the on-station orbit (determined by inclination and ascending node position) for a free-drift station-keeping strategy is required to be at a certain position in *inertial space*. Owing to the rotation of the Earth, the Earth-fixed transfer orbit ascending node and the inertially-fixed on-station orbit ascending node will coincide once every 24 hours.

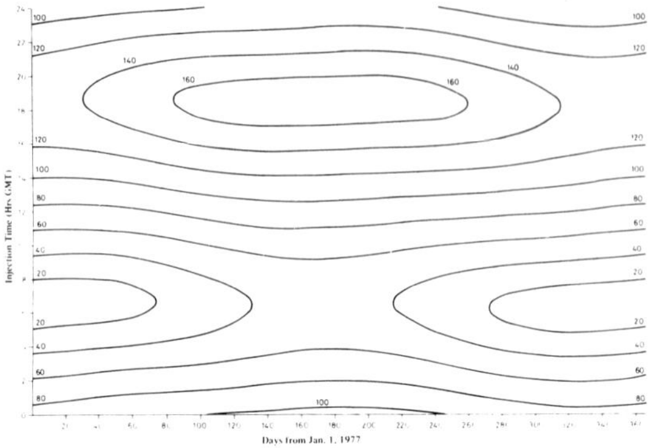

Fig.2.4 Solar Aspect Angle Contours – Injection Attitude

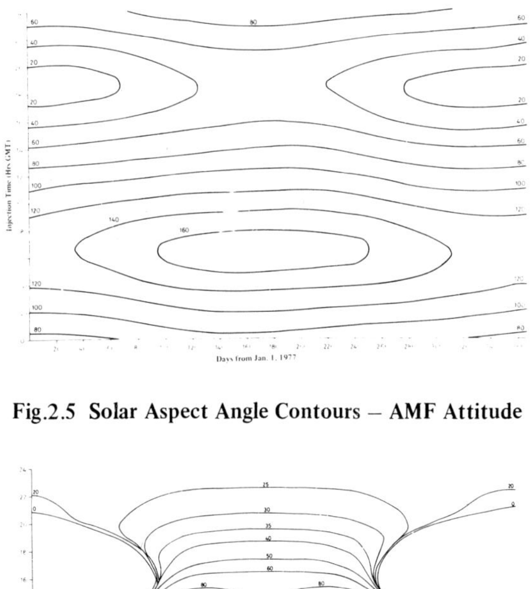

Fig.2.5 Solar Aspect Angle Contours – AMF Attitude

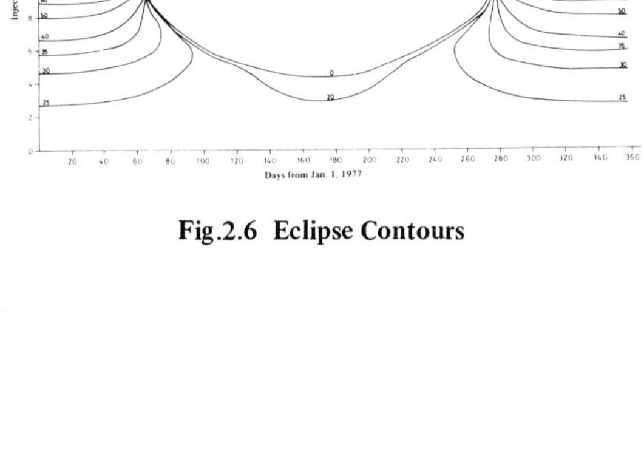

Fig.2.6 Eclipse Contours

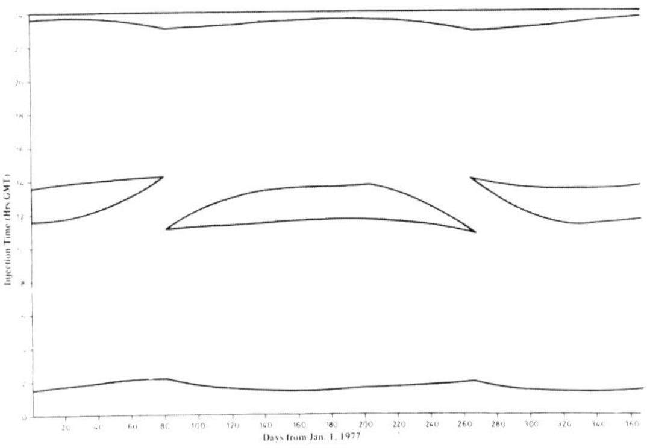

Fig.2.7 Launch Window Due to Solar Aspect Angle and Eclipse Constraints

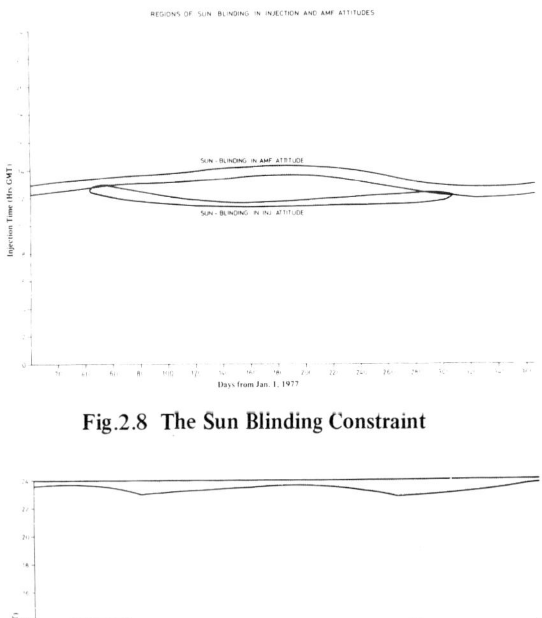

Fig.2.8 The Sun Blinding Constraint

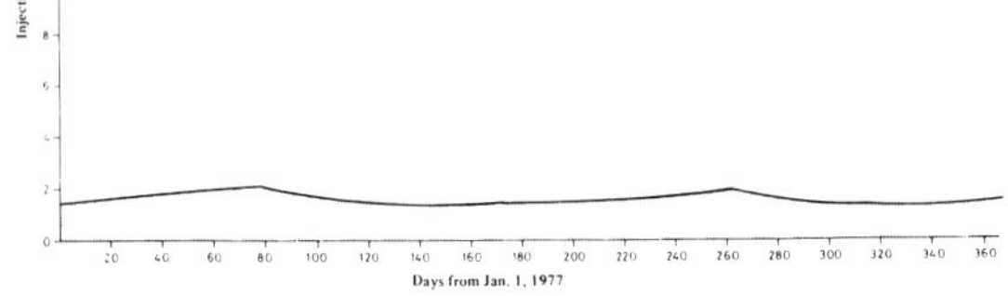

Fig.2.9 Final Launch Window

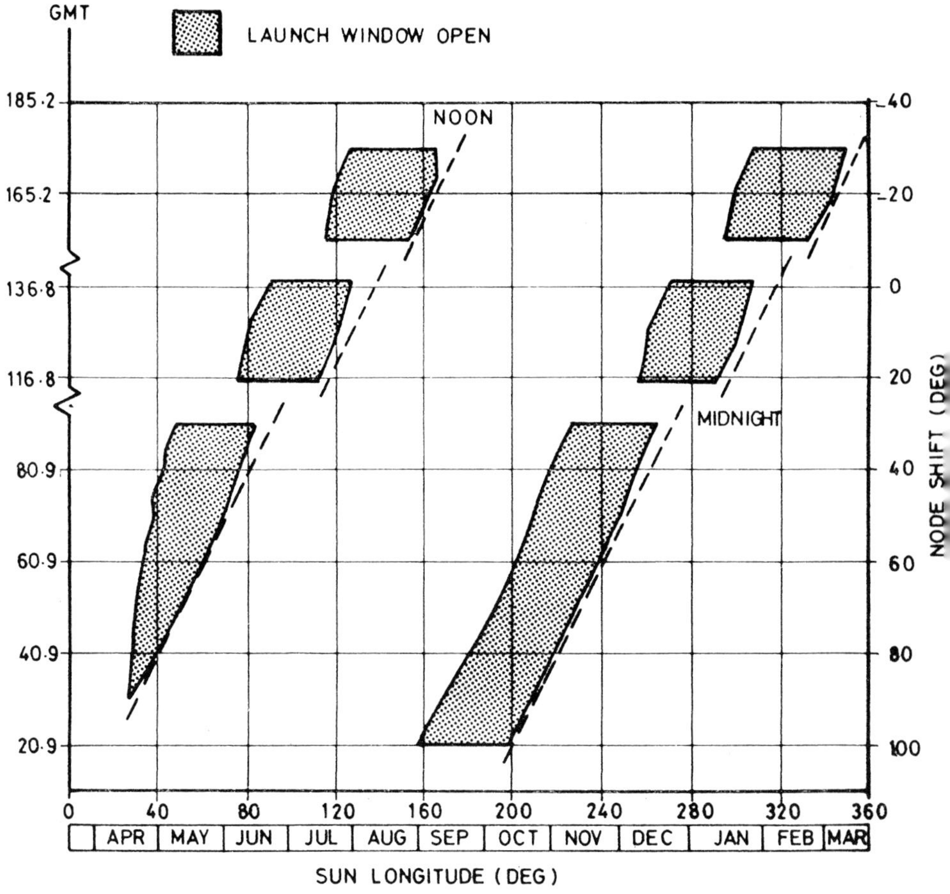

Fig.2.10 MARECS Launch Window

Hence if no node-shifting is available, there will be only one launch opportunity (and a brief one) every 24 hours. If, however, this brief launch slot is precluded due to the violation of any one of the several other launch window constraints, it becomes evident that launch opportunities will be severely restricted. To open the launch window we need to expand the range of allowed positions in inertial space of either the transfer orbit or drift orbit ascending nodes. As the drift orbit node is moved away from its optimum position, the period of on-station free-drift within the inclination limit reduces, and eventually results in the need for some North-South station-keeping. Obviously this is inconsistent with the concept of a free-drift mission and therefore any variation in the drift orbit node must be restricted, to exclude the necessity for station-keeping. Section 6.3.3.2 describes the strategy for obtaining the range of acceptable drift orbit node positions in more detail. The variation in drift orbit node position obtained in this way, however, may be very limited and will not necessarily result in a significant expansion of the launch slot. If, on the other hand, the ascending node of the transfer orbit, in inertial space, is allowed to vary, then both the number of launch opportunities and the duration of each launch slot may be increased until the required overall launch window is achieved. When this is done, we require the capability of changing the ascending node from any of the transfer orbit values to one or all of the drift orbit values dictated by the free-drift strategy. Hence node-shifting results in opening the launch window. Figure 2.11 illustrates the above argument.

Analysis has shown that a node-shift capability of about $\pm 80^{\circ}$ is usually required to satisfy launch window requirements. The node-shifting manoeuvres are carried out either by the apogee motor or by the satellite's reaction control system, or by a combination of the two.

For node-shifting to be performed efficiently, the elements of the transfer orbit must be selected to suit the node-shift required. The transfer orbit elements are, however, achieved by the guidance program stored in the launcher's guidance system, and it is totally impractical to provide for a different program for each node-shift in the range -80° to $+80^{\circ}$. The range is therefore divided into a small subsets of ranges, typically 2 or 3, and the transfer orbit elements are designed to cater for each of the ranges with acceptable efficiency. We therefore have 2 or 3 guidance programs for the launcher. Division of the total node-shift range into subset is done on the basis of trial and error, guided by experience and an understanding of the physical nature of the problem. Optimum transfer orbits are derived, which maximise the satellite mass on-station (see Section 3.2). These orbits are then used to determine the launch window. If this is found to be inadequate, the node-shift ranges are adjusted, and the calculation re-done, until satisfactory results are obtained.

To facilitate the launch window assessment, computer programs are prepared. These evaluate the main constraints, i.e. solar aspect angle and eclipse duration, for the required period, and indicate whether launch is possible or not. Subsidiary constraints such as Sun-blinding and lunar eclipses are evaluated separately and their results super-imposed on the above. The overall launch window is then obtained.

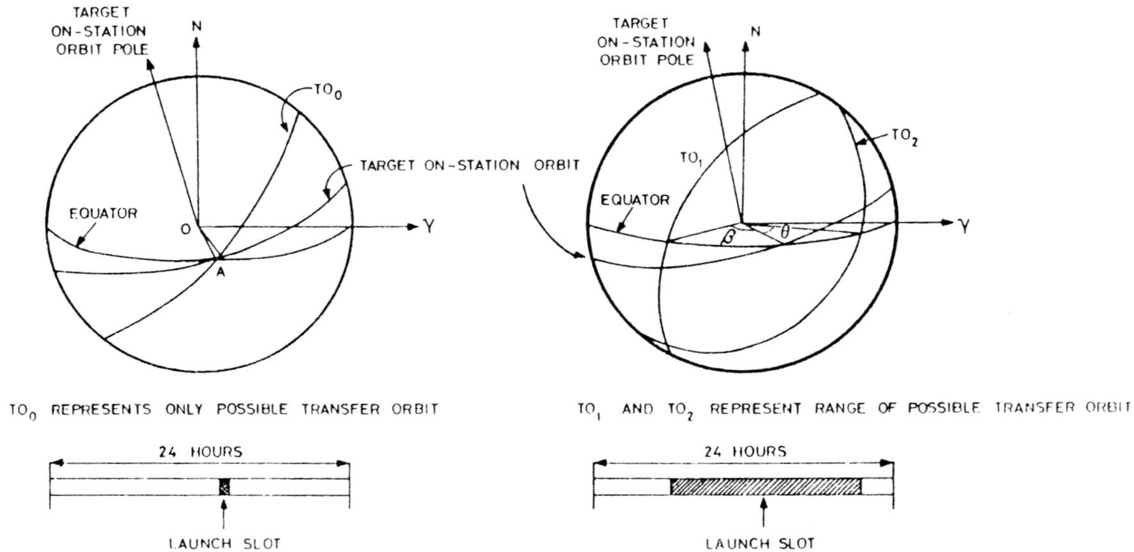

Zero Node-Shift Capability | $\beta + \theta$ **Node-Shift Capability**

Fig.2.11 Node Shift Strategy

2.2.3 Launch Window for Shuttle Launch

When the launch vehicle is the *Shuttle*, slightly different considerations apply for the launch window. These are outlined briefly here.

Unlike conventional launch vehicles, for a *Shuttle* launch, the time of injection into transfer orbit is not closely tied to the time of launch. This is because there is flexibility in the selection of both deployment time and perigee stage ignition time, once the *Shuttle* has achieved the near-Earth parking orbit. Injection opportunities will occur every time the parking orbit crosses the Equatorial plane. Owing to the regression of nodes due to the Earth's oblateness, the locations of the injection opportunities move in inertial space. For the *Shuttle* parking orbit, the regression of nodes is about 7.5^{o}/day. To allow some scheduling flexibility, it will be necessary to select a small number of injection opportunities for the mission, rather than just one. Once the injection opportunities have been selected, the range of spin axis attitudes is determined. The launch window is then evaluated to satisfy all the constraints for this range of spin axis attitudes.

Owing to the requirement to mix payloads on any flight, spacecraft requiring launch into geostationary orbits have to comply with a set of rules laid down by NASA. At the present time these are not firm and fixed, more in the nature of guidelines, but it is clear that further in the future fairly strict rules will be imposed.

NASA has defined the term *standard launch window* to mean, for shared flights from Kennedy Space Centre to the standard mission orbit of 160 nautical miles circular orbit at 28.5^{o} inclination (nominal):

i) launch on any date, selected by NASA, from the following periods:

   Non-Standby Payloads: The launch period shall begin on the planned launch date and end 90 days later. Once the firm launch date has been set, the launch period shall begin on that date and end 60 days later.

   Standby Payloads: The launch period shall encompass the entire pre-negotiated 1-year launch period.

ii) launch at any time of day, selected by NASA, from either of two launch windows, one centred on 12:23 GMT and the other centred on 00:23 GMT. The total width of each of these launch windows shall be 2 hours.

To satisfy the above requirements the user must select injection opportunities consistent with the requirements below:

- All on-orbit mission operations shall be capable of being performed within a 24-hour period of on-orbit time without conflicting with the performance of routine crew activity during this period.

- For launch during at least 80% of each of the appropriate *standard launch windows* for shared flights, payloads shall be designed such that all orbit/ Sun constraints are satisfied for both ascending and descending equatorial node deployments on the 33rd orbital revolution of the Shuttle. The user may, however, specify other constraints which preclude an actual 33rd-revolution deployment.

- Shared flight payloads shall be capable of being deployed on three consecutive orbital revolutions during at least two periods separated by at least 12 hours and occurring between the 5th and 56th orbital revolutions after launch.

- In order to provide compatibility with crew work and rest periods, if ascending and descending node injections are not both available on any launch (e.g. if noon GMT launch means ascending nodes only), then two longitude bands, both at least 25° wide, must be provided. One will be in the range of 40°E to 165°E, for ascending node injections, and one in the range of 15°W to 140°W for descending node injection. If ascending and descending node injections are both available on any launch, then a single longitude band, at least 25° wide, must be provided and it may lie in either of the above stated ranges.

The above requires that the short-listed injection opportunities be carefully analysed to ensure that:

- All spacecraft constraints, such as thermal- and power-related constraints, are satisfied.

- Adequate ground station coverage is obtained for all operational requirements in parking and transfer orbit.

The injection opportunity for a specific payload will be selected by NASA from the supplied short-list, bearing in mind the requirements of the other payloads and Shuttle-related constraints.

2.2.4 Launch Window for Ariane Launch

During the early years of the *Ariane* programme, when the operational vehicle was *Ariane 1*, single payloads were generally launched. The vehicle lift-off time was then constrained by considerations deriving from the spacecraft's systems. In other words, the launch window was specified *by the spacecraft*, and the only launch vehicle constraint that applied was the duration of a launch opportunity. This was required to be 45 minutes, to allow some flexibility in the count-down procedure.

With the advent of the *Ariane 4* vehicle this situation will change. *Ariane 4* is a large vehicle capable of injecting multiple payloads into GTO. To ensure mixability of the payloads, and to ease the launch agency's payload manifesting task, a launch window will be *imposed upon all spacecraft* desiring injection into GTO. The launch opportunity duration will be 45 minutes, and only a midnight launch will be available. This launch window is called the *standard launch window* and prospective payloads will be required to conform to it. It is illustrated in Figure 2.12, and has been derived by Arianespace after due consideration of the launch window constraints applicable to both spin-stabilised and three-axis-stabilised spacecraft. Conforming to this launch window should not, therefore, pose undue problems.

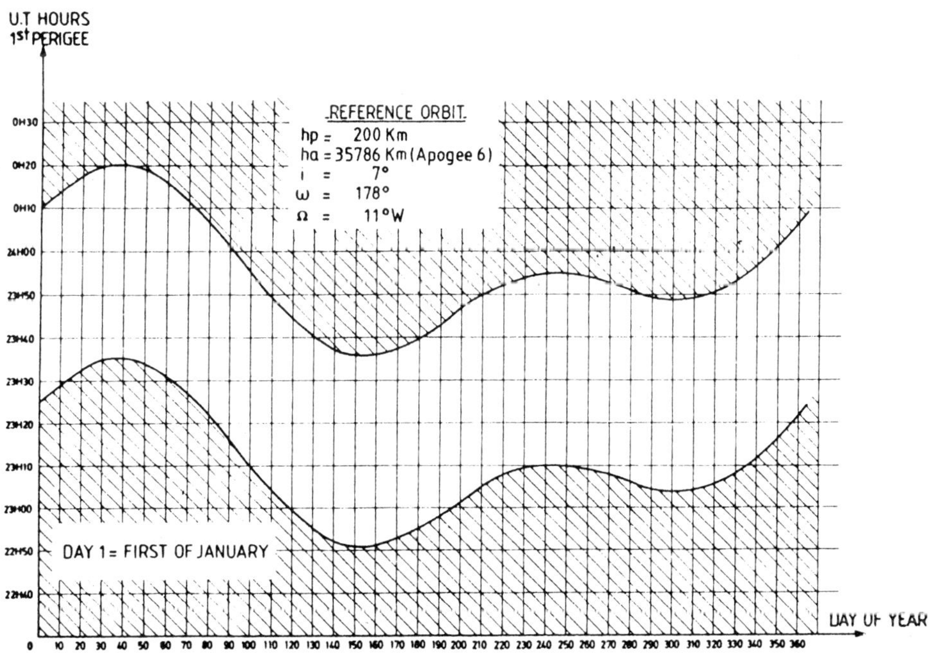

Fig.2.12 Ariane "Standard Launch Window"
(Reproduced by permission of Arianespace)

2.3 The Launch Sequence

In order to define the mission during the launch phase, the launch event sequence is compiled. The launch event sequence is mostly out of the control of the satellite mission planner in so far as it is determined essentially by the selection of the launch vehicle and transfer orbit characteristics. Once the launcher is selected, the event sequence will be available from the vehicle's User Manual. Small variations can be expected due to the specific trajectory that will be flown by the vehicle to place the payload into the desired transfer orbit. Other mission-related differences will be the payload separation events, specially if a dual-launch (i.e. the same launcher putting into orbit two satellites) is being considered.

An example of the launch event sequence – that for MARECS launched on *Ariane* – is shown in Table 2.3.

Increasing use of the *Shuttle* will require careful consideration of the payload deployment sequence. If several payloads are being launched on the same *Shuttle* flight, careful control of mutual separations between the various payloads will be important if interference and collisions are to be avoided. Active or passive means to achieve the required separation have to be considered. Active separation can be achieved by actuation of satellite on-board thrusters, while passive separation is achieved through differential drag effects. Passive separation can be achieved also by timing the deployment from the Orbiter's cargo bay.

Table 2.3

The Launch Event Sequence for MARECS on Ariane

Event	Time (Secs)
First Stage Ignition	-3.0
Lift Off	0.0
Umbilical Eject	1.0
Start of Pitch Program	20.0
First Stage Burn Out	144.0
First Stage Separation	149.0
Second Stage Full Thrust	182.0
Fairing Jettison	243.0
Second Stage Burn Out	284.0
Second Stage Separation	289.0
Third Stage Full Thrust	299.0
Third Stage Cut-Off by Guidance Computer	856.0
Start of Payload Re-orientation to Nominal AMF Attitude	858.0
Slow Spin-Up to 10 rpm.	
Latest Time of Payload Separation	1036.0
Third Stage Tumble – Valves Opened	1096.0
End of Ariane Mission	

2.4 The Injection Error Covariance Matrix

Although the injection error covariance matrix is not used explicitly in any of the mission analysis activities associated with the launch phase, it is discussed in this section because it is descriptive of the injection accuracy of the launch vehicle.

Dispersions in the ascent trajectory of the launch vehicle will result in errors in the transfer orbit parameters. As there is coupling between the elements of the error vector, we have a fully populated error covariance matrix where diagonal terms are the variances of the individual elements. The injection error covariance matrix can be expressed in many forms. For example, the matrix can be expressed in terms of the elements of the transfer orbit:

a	=	semi-major axis
e	=	eccentricity
i	=	inclination
ω	=	argument of perigee
Ω	=	right ascension of ascending node (relative to the Greenwich meridian)
f	=	true anomaly

Or it can be expressed in terms of satellite position and velocity at injection:

r	=	radius
long.	=	geographic longitude at injection
lat.	=	geographic latitude at injection
V	=	velocity
Az.	=	azimuth angle of spin axis
El.	=	elevation angle of spin axis

Knowledge of the covariance matrix is required at an early stage in mission design, as it is used to derive the propellant required for the apogee manoeuvre and station acquisition.

2.5 Further Reading

The information on launch vehicles, required for mission design, is obtained from the launch vehicle User Manuals. Examples of these are cited below.

(a) PAM-D/PAM-DII User Requirements Document.
McDonnell Douglas Astronautics Company, MDC G6626F and MDC H0421A, July 1985.

(b) Ariane 2 and 3 User's Manual.
Arianespace; Issue 1, Revision 4.

(c) Ariane 4 User's Manual.
Arianespace; Issue 1, Revision 2, January 1986.

(d) Space Shuttle System Payload Accommodations.
NASA JSC 07700, Vol.XIV, Revision H, May 1983.

General information on launch vehicles can be found in a wide variety of aerospace magazines and in technical journals. Examples are:

(e) Kidger, N. : 1982, "Japanese Space Plans for 1982-1985", *Spaceflight*, Vol.24, May, pp.202-204.

(f) Velupillai, D. : 1984, "Commercial Rockets", *Flight International*, 14 January, pp.95-105.

(g) Clark, P.S. : 1982, "Soviet Launch Vehicles: An Overview", *J. of British Interplanetary Soc.*, Vol.35, No.2, Feb., pp.51-58.

(h) Clark, P.S. : 1977, "The Proton Launch Vehicle", *Spaceflight*, Vol.9, Sept., pp.330-333 and 340.

(i) Parfitt, J. : 1985, "The Soviet Proton Launcher", *Space Markets*, No.2, Summer 1986.

A discussion of the launch window derivation for STS-launched spacecraft is presented in the paper below:

(j) Wright, W.H. : 1979, "Launch Window for a Shuttle-Based Geosynchronous Orbit Mission", *J. of Spacecraft*, Vol.16, No.6, pp.439-441.

CHAPTER 3

TRANSFER ORBIT

3.1 Introduction

The function of the transfer orbit is to take the spacecraft up to geosynchronous altitude for the purpose of performing the apogee manoeuvre. The transfer orbit is the intermediate stage in a *Hohman Transfer* manoeuvre between the near-Earth, circular, parking orbit and the geosynchronous orbit, and as such, one apse is at parking orbit altitude, while the other is at geosynchronous orbit altitude. This gives the transfer orbit a high eccentricity, about 0.73. Since, in general, the parking orbit inclination is non-zero and the geosynchronous orbit inclination is near-zero, a plane-change manoeuvre is also required. To perform this efficiently, it is executed at the apogee of the transfer orbit, leading to the requirement that the apogee be also a node of the transfer orbit. The perigee height is chosen to be the minimum required to overcome the adverse effects of aerodynamic drag on the orbit and kinetic heating on the satellite.

These requirements, therefore, establish the main characteristics of the transfer orbit. Table 3.1 shows the transfer orbit characteristics of the *Delta*, *Ariane* and *STS/PAM-D* vehicles. It should be noted that the inclination of the orbit is closely related to the latitude of the launch site, and that for maximum efficiency consistent with range safety requirements, a launch azimuth in the region of 90° is selected.

Though the launch vehicle governs the main characteristics of the transfer orbits, small variations of the elements are possible, and the exact orbit elements can sometimes be selected by the mission designer based upon the requirements of the spacecraft. The elements that are usually varied are:-

- apogee height
- inclination
- perigee position.

The basis upon which the selection is made is mission-dependent, but generally involves optimising the transfer orbit to achieve maximum spacecraft mass on station. The following section describes the optimisation of the transfer orbit and the various considerations that arise.

With the increasing use of multiple-payload launch vehicles, the characteristics of the transfer orbit may not be available for optimisation to suit a particular payload. The transfer orbit selected by the launch vehicle authority is such as to be suitable for all the payloads being carried, and cannot excessively cater for one payload to the detriment of the other(s). This constraint must be recognised by the spacecraft's mission designer at an early stage.

Figure 3.1 illustrates the main characteristics of the transfer orbit.

Table 3.1

Transfer Orbit Characteristics

		Delta	**Ariane**	**STS/PAM-D**
Apogee height	(km)	35786.	35786.	35786.
Perigee height	(km)	185.	200.	296.
Inclination	(deg)	28.7	7.0	27.3
Eccentricity		0.73	0.73	
Argument of perigee	(deg)	180.0	180.0	Selectable
RA of ascending node	(^{o}E of GM)	173.0	163.2	Selectable
Launch site latitude	(^{o}N)	28.45	5.24	28.45
Launch azimuth	(deg)	95.0	90.0	

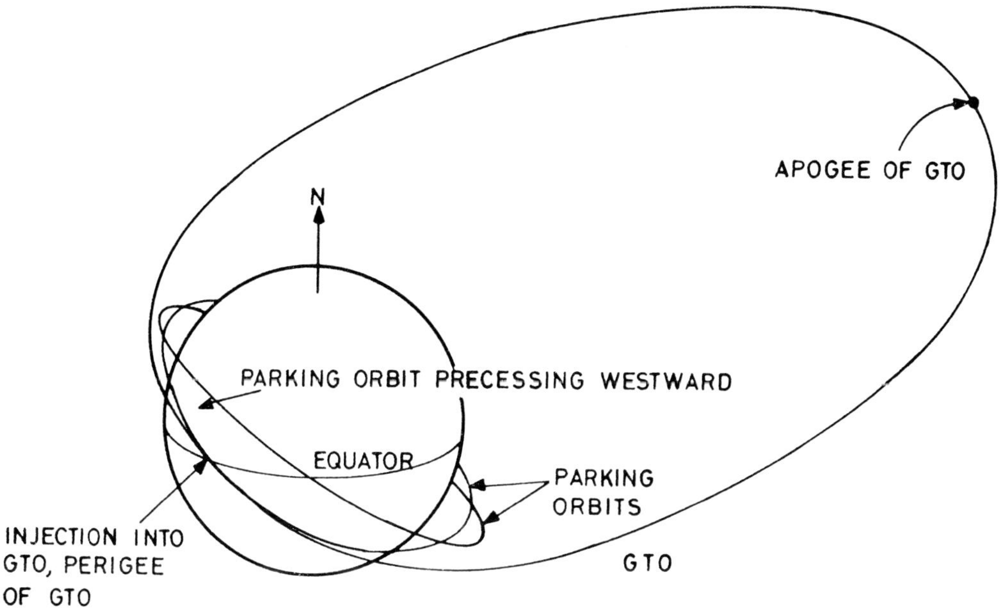

Fig.3.1 Characteristics of the Transfer Orbit

3.2 Orbit Optimisation for Spacecraft with Solid Propellant Apogee Motors

The purpose of transfer orbit optimisation is to achieve the maximum possible spacecraft mass at the beginning of its useful life on station. The incentive to do this is provided by the very high cost of launching a spacecraft into geostationary orbit. The launch cost is so high as to be comparable with the cost of the spacecraft itself. Orbit optimisation maximises the spacecraft's useful mass by varying certain orbit parameters within allowable constraints.

Small changes in certain transfer orbit elements result in significant changes to the payload capability of the launch vehicle. For the *Ariane 1* launch vehicle, for example, the payload mass is reduced by 0.02 kg for each kilometre that the transfer orbit apogee height is increased above the nominal. The equivalent reduction for an increase in perigee height is 0.7 kg per kilometre. Similar trade–off factors relate the payload mass to orbit inclination and argument of perigee. Since the spacecraft transits from the transfer orbit to the geostationary orbit through the apogee manoeuvre and the station acquisition manoeuvres, changes in transfer orbit elements also result in changes in the propellant required to execute the apogee and station acquisition manoeuvres. These must also, therefore, be considered in the optimisation process.

Also to be considered in the optimisation process is the non-nominal performance of the launch vehicle. To allow for transfer orbit dispersions, the optimisation is carried out on a statistical basis, and the spacecraft beginning-of-life (BOL) mass is optimised taking into account non-nominal performances of the launch vehicle and spacecraft propulsion system to some probability level – usually 99%.

Let us now see, in a simple way, how small changes in orbit parameters can lead to some increase in spacecraft BOL mass. Let the nominal mission scenario be as follows:

- the spacecraft is injected into a nominal transfer orbit whose apogee is at geosynchronous altitude;
- the apogee motor is fired at an apogee whose geographic longitude is due West of the spacecraft's operational longitude;
- after the apogee motor firing, manoeuvres are performed to set up an Easterly drift of the spacecraft towards the operational longitude;
- when the operational longitude is reached, a further manoeuvre is performed to stop the spacecraft at the operational longitude (thereby achieving the geostationary orbit).

Consider now a transfer orbit apogee height slightly lower than nominal. Since this is a lower-energy orbit than the nominal one, the launch vehicle can launch a slightly heavier payload than nominal. Now target the apogee manoeuvre to achieve a post-burn orbit in the equatorial plane, with the opposite apse at geostationary height (the manoeuvre is itself made at an apse which is lower than synchronous height). This manoeuvre requires somewhat less energy than the nominal manoeuvre, and hence requires less propellant. After the burn, the spacecraft will be in a slightly lower orbit than geostationary orbit, and an Easterly drift rate will automatically be established. This saves further manoeuvre propellant compared to the nominal case. When the spacecraft reaches its operational longitude, a manoeuvre is performed to circularise the orbit and so stop the spacecraft at that longitude. It can be seen that when compared with the nominal, spacecraft mass gains have occurred due to:

- greater launcher payload capability;
- smaller apogee motor propellant mass required;
- smaller station acquisition propellant mass required.

A saving in spacecraft on-board fuel requirement can be translated into a gain in the useful mass the spacecraft can carry for the same on-orbit lifetime. Targeting the transfer orbit apogee in such a way as to obtain a uni-directional drift-rate after the apogee burn is known as *apogee biasing* and is a much-used technique for increasing the spacecraft's BOL mass.

Another technique which is used is that of removing a very small amount (between 1 to 2 degrees) of the parking orbit inclination by the third stage of the launch vehicle. The velocity increment, ΔV, required for inclination change is directly dependent upon the velocity of the spacecraft along the orbit when the manoeuvre is made. It is very much more efficient, therefore, to remove the inclination when the spacecraft is at transfer orbit apogee and travelling slowly, than when the spacecraft is at perigee and travelling very much more quickly. It would seem perverse, therefore, to remove inclination at the perigee of the transfer orbit and indeed only a very small amount of inclination is removed at perigee during injection into GTO. Two factors make it reasonable to do this, firstly, if more propellant is expended in injecting the spacecraft into transfer orbit the spacecraft mass injected into transfer orbit is lower. Secondly, if the propulsion efficiency of the launch vehicle final stage is higher than that of the spacecraft's apogee motor, it is more efficient to allocate to it a greater burden of the propulsive task.

For *Delta* and *PAM-D* launches, removing 1.5^{o} of inclination at injection increases the spacecraft BOL mass by about 2 kg. This may not seem much, but it is achieved at no cost, and is, therefore, well worth having.

When both apogee biasing and inclination removal at injection are used together, a worthwhile gain in spacecraft useful mass is achieved. The optimisation is performed by computer, on the following parameters:

- transfer orbit apogee height,
- transfer orbit inclination,
- transfer orbit argument of perigee,
- apogee boost motor propellant mass.

The optimisation algorithm takes account of operational constraints, such as the requirement to make provision for an opportunity for a back-up manoeuvre in addition to the prime one. It includes an operationally viable apogee motor firing strategy and a station-acquisition strategy.

A further effect to be considered is the influence of non-nominal launch vehicle performance. The spacecraft design is usually required to deal with launch vehicle performance variation to the level of 2.58 standard deviations of the nominal. This covers 99% of all cases that can arise. Since the spacecraft's propulsion system must compensate for injection dispersions, a variation of the spacecraft BOL mass results. It is the function of orbit optimisation to maximise the lowest value of BOL mass due to launch vehicle dispersions. The largest dispersion caused by non-nominal vehicle performance is on apogee height. It is, therefore, this parameter that is the subject of the optimisation. Clearly, this adds a further factor to the apogee biasing discussed earlier.

3.2.1 Optimising the Transfer Orbit for a Free-Drift Mission

Some geostationary spacecraft do not require to be kept within tight latitude limits, though, in order to avoid interference with other spacecraft, tight longitude limits must still be observed. As the latitude deadband is wide, it may be possible to avoid North–South station-keeping altogether, but if this is to be the case, the orientation of the orbit plane at the start of the spacecraft's useful life on station must be carefully set up. The orientation of the orbit plane is defined by the position of the orbit normal, known as the orbit pole.

Owing to the effects of Luni-Solar perturbations, the orbit pole tends to drift towards the Vernal Equinox for small inclinations. This is shown in Figure 3.2. Now if a spacecraft wishes to avoid North–South station-keeping manoeuvres, and still wishes to minimise its orbit inclination over its operational lifetime, it is clear that the orbit pole must initially be placed on the side of the North Pole, remote from the Vernal Equinox. Only then will the orbit inclination first reduce, pass through a minimum, and then increase again. This implies that the initial value of the right ascension of the orbit's ascending node must be in the region of 270°.

As mentioned previously (Chapter 2, Section 2.2.2, and Figure 2.11), if the launch window is to be rendered sufficiently open as to allow a launch on most days of the year, some node shifting must be carried out as part of the apogee manoeuvre. To transfer between two orbits of different inclinations, such as the transfer orbit and the geosynchronous orbit, a manoeuvre is carried out at their point of intersection. If this point of intersection is not in the Equatorial Plane, the right ascensions of the ascending nodes of the two orbits will be different. A manoeuvre carried out at this point will therefore result in a node shift, and the node shift will increase as the manoeuvre latitude increases. The node shift geometry is shown in Figure 3.3.

.Let us now consider the apogee manoeuvre. For maximum efficiency, the manoeuvre must be performed exactly at apogee. To achieve a given node shift, therefore, the apogee position must be at the appropriate latitude. However, launch window considerations dictate that the spacecraft must be capable of carrying out a range of node shifts, the exact node shift required being dependent on the time and day of launch. It is clear, therefore, that if each node shift is to be performed with maximum efficiency, the position of the apogee becomes a variable, determined by the time of launch. If this is not possible, and the apogee position is fixed, then all node-shifts bar one are achieved non-optimally.

Conventional expendable launch vehicles do not allow the apogee position to be targeted by dependence on launch time. At best, a discrete number of apogee positions (usually 2 or 3) is available, corresponding to launch seasons. In such a case it is clear that the total required node shift range must be sub-divided into sub-ranges, each of which is satisfied by a particular set of transfer orbit parameters. Selection of a launch season, then, implies selection of a node shift range, which in turn dictates the transfer orbit parameters. If the launch season is missed, the next available one could well require a change in the node shift range and the transfer orbit parameters.

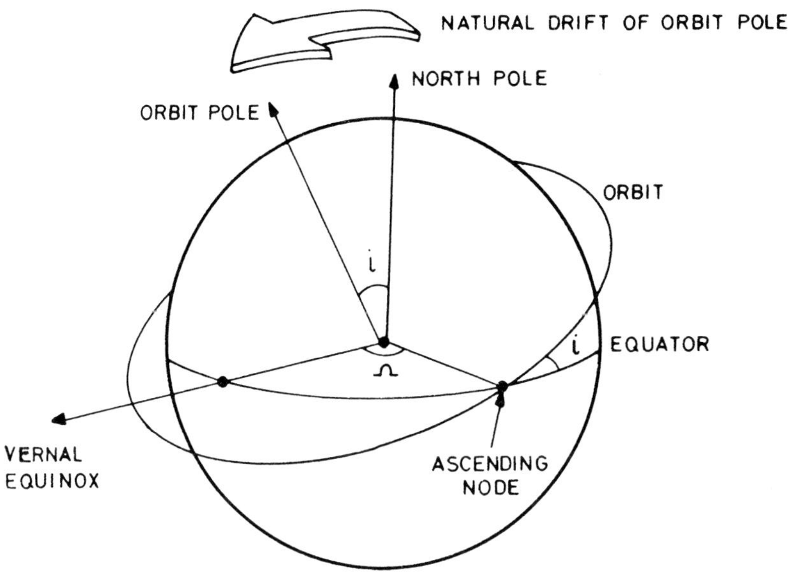

Fig.3.2 GEO Pole Drift due to Luni-Solar Perturbations

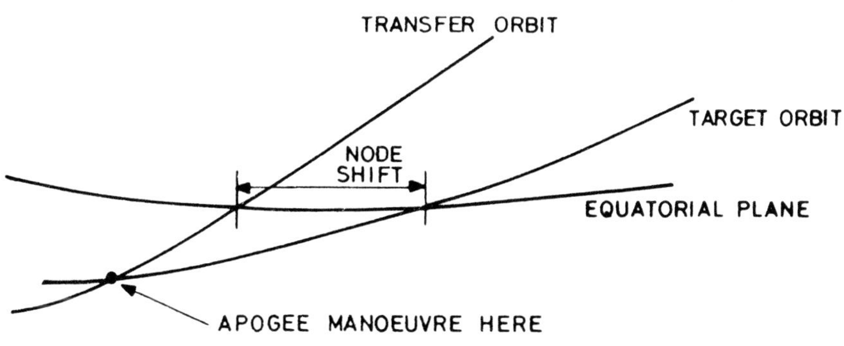

Fig.3.3 Node Shifting by Apogee Manoeuvre

The situation with the *Shuttle* is somewhat different, and a degree of retargeting of transfer orbit elements as a function of launch time is possible. The presence of *man-in-the-loop* allows the deployment time of the spacecraft-plus-upper stage stack to be varied as a function of launch time. The deployment time and deployment attitude can both be determined in parking orbit, either by the Orbiter crew consulting look-up tables, or by the information being telemetred to the Orbiter crew from the control centre on the ground. The apogee position can therefore be closely tailored to the node shift required.

Let us now consider the transfer orbit optimisation problem. It is clear that the argument of perigee, which determines the apogee position, becomes an important optimisation parameter. Further, since the velocity increment required at apogee is a function of node shift, the size of the spacecraft's apogee motor, which must remain fixed once fabricated, is also an important optimisation parameter.

The Optimisation Procedure for Conventional Launch Vehicles

The first step is to identify the node shift range necessitated by launch window considerations. This range is then sub-divided into two or three sub-ranges, having due regard for launch vehicle performance. Since each sub-range is associated with a unique set of transfer orbit elements, the optimisation is, initially, carried out separately for each sub-range. The variables in the optimisation are:

- apogee height,
- inclination,
- argument of perigee,
- apogee motor propellant mass.

These parameters are varied to maximise the lowest BOL mass within the range of node shifts being considered. To ensure good resolution, the BOL mass is calculated for between 5 to 10 individual node shifts within the sub-range being considered. Both prime and back-up apogee manoeuvre opportunities are given equal weight, and the precession of the argument of perigee due to the Earth's oblateness is taken into account.

This procedure results in as many optimal transfer orbits, apogee motor sizes, and BOL masses, as node shift sub-ranges. At this point, examination of the BOL masses indicates whether the initial node shift ranges were well chosen. If the variation in the BOL masses associated with the optimised orbits is greater than about 15 kg, the node shift sub-ranges should be revised, and the procedure repeated until the variation in BOL mass is 10 kg or less. Once this has been achieved, the next step is taken.

It was mentioned above that each sub-range had associated with it an optimal apogee motor size. The spacecraft can have but a single apogee motor. Of the optimal sizes, the one that is selected is that which is associated with the lowest BOL mass. The motor size is then kept fixed, and the sub-ranges re-optimised using the same procedure as before. This results in a set of optimal transfer orbits for the node-shift sub-ranges considered.

The Optimisation Procedure for the Shuttle

Let us now consider the optimisation for the *Shuttle*. Here the argument of perigee becomes a free parameter that is chosen to suit the node shift required. The optimisation parameters are:

- apogee height,
- inclination,
- apogee motor propellant mass.

Owing to the characteristics of the launch system, and the fact that the argument of perigee is no longer an optimisation parameter, the node shift range does not need to be sub-divided. The procedure is the same as the one outlined above and results in a set of optimal transfer orbit elements and optimal apogee motor size.

It must be recognised that a particular mission may have constraints other than those mentioned here. These must, as far as possible, be reflected in the choice of the optimisation variables, and in the procedure used to calculate the spacecraft BOL mass. Some constraints that could arise are, the size, and size variation of the apogee motor, the station acquisition strategy, the apogee motor firing strategy, and the launch window. Various numerical methods may be employed in the optimisation. Since the number of variables is small, the optimisation task is not a difficult one.

The Optimisation Procedure for MARECS

As an illustrative example, we shall consider the transfer orbit optimisation for ESA's MARECS spacecraft launched by *Ariane*. The MARECS spacecraft drifts freely in latitude while being maintained within strict longitude limits. At the start of its operational life the spacecraft is placed at an orbit inclination of 3° in such a way that the inclination first decreases, passes through a minimum, and then increases again in the latter half of its life. In this way the spacecraft remains within latitude limits of $\pm 3^o$ throughout its 7-year operational lifetime.

The *Ariane* launch vehicle allowed the spacecraft a maximum of three nominal transfer orbits with which to open the launch window. The mission design task was to determine the node shift range to be covered by each transfer orbit, and to optimise the three transfer orbits in such a way as to maximise the spacecraft's BOL mass.

Owing to the performance characteristics of the *Ariane* launch vehicle, the node-shift range assigned to each transfer orbit is substantially unequal. For MARECS, the following optimal ranges were arrived at:

(a) $35^o - 95^o$

(b) $4^o - 16^o$

(c) $-25^o - -15^o$

With these node-shifts, the MARECS launch window is as shown in Figure 2.10 in Chapter 2. This affords a launch on 325 days of the year.

The initial optimisation of the transfer orbits considered the following optimisation parameters:

- apogee height,
- inclination,
- argument of perigee,
- apogee boost motor propellant mass.

The results are shown in Table 3.2. It can be seen that for the three transfer orbits, the spacecraft BOL mass and the apogee boost motor propellant mass lie within acceptably narrow bands. This confirms that the choice of node-shift ranges was a good one. The lowest BOL mass and motor propellant mass occur for the transfer orbit with negative node-shifts. We now fix the apogee motor propellant mass at this (smallest) value and re-optimise the other two transfer orbits with this value of propellant mass. The results are shown in Table 3.3. It can be seen that the spacecraft BOL mass now falls within an even smaller range. This is a satisfactory result, and indicates that the three transfer orbits will give the highest BOL mass under the given conditions of required node shift range and fixed apogee boost motor propellant load.

3.3 Orbit Optimisation for Spacecraft with Liquid Propellant Apogee Engines

Though the main ideas behind the transfer orbit optimisation remain the same, some differences do occur when the spacecraft's apogee engine is of the liquid propellant type. This is due to the following main characteristics of the apogee engine:

- the burn size is adjustable on orbit by selection of engine on/off times;
- the burn is non-impulsive;
- the engine is re-startable, and hence more than one burn may be employed for the apogee manoeuvre.

Table 3.2

Initial Optimal Transfer Orbits for MARECS

Transfer Orbit	1	2	3
Node-Shift Range (deg)	35 – 95	4 – 16	-25 – -15
Inclination (deg)	7.5	8.8	11.1
Argument of Perigee (deg)	157.6	173.3	185.3
Apogee Height (km)	42327.7	42375.8	42431.4
ABM Propellant (kg)	681.3	684.3	680.5
Minimum BOL Mass (kg)	990.8	995.8	985.5

Table 3.3

Final Optimal Transfer Orbits for MARECS

Transfer Orbit	1	2	3
Node-Shift Range (deg)	35 – 95	4 – 16	-25 – -15
Inclination (deg)	8.1	10.5	11.1
Argument of Perigee (deg)	158.5	174.2	185.3
Apogee Height (km)	42175.0	42376.7	42431.4
ABM Propellant (kg)	680.5	680.5	680.5
Minimum BOL Mass (kg)	984.2	984.1	985.5

Considering a single burn manoeuvre first, the overall mission strategy requires the thrust vector to be steered optimally to achieve minimal fuel use. Incorporation of the steering optimisation within the transfer orbit element optimisation is not practicable computationally. Hence a simplified method is required. The method adopted consists of:

- assuming that the apogee burn is impulsive as far as orbit mechanics is concerned;
- assuming some manoeuvre inefficiency to take account of the fuel penalty due to the non-impulsive nature of the burn;
- adopting a burn strategy that achieves the required plane change, and adjusts the burn magnitude and direction to minimise the subsequent station acquisition velocity increment;
- in other respects using the same methodology as in the case of a satellite with a solid propellant apogee motor.

This leads to a definition of the transfer orbit elements as before. We cannot, however, stop here, because the above procedure takes no account of the inherent tendency of the long duration burn to raise the apogee height. The final step in the optimisation procedure, therefore, involves a return to reality by treating the apogee manoeuvre as a long duration burn, with the appropriate thrust vector steering law. We start with the optimum transfer orbit elements derived above, and parametrically reduce the apogee height. At each step we optimise the long duration burn to take the spacecraft from the transfer orbit to the target drift orbit derived by the above optimisation procedure. We take into account the variation in the launcher's payload performance with apogee height, and calculate the mass of the spacecraft into transfer orbit. Finally, we select the apogee height that yields the maximum spacecraft mass into drift orbit. Thus the optimal transfer orbit is fully defined.

An exercise such as the above was carried out for the OLYMPUS spacecraft, and it was found that the apogee height had to be lowered by about 200 km to yield optimal results.

We consider now the case of the multi-burn apogee engine firing (AEF) strategy. The differences from the solid motor case are somewhat greater. Here the apogee manoeuvre algorithm is completely different. The multi-burn apogee manoeuvre is discussed in Section 4.3.1.2, and this section should be read, before proceeding further, to obtain an understanding of the reasoning behind such a strategy.

Once a multi-burn strategy has been selected with due regard to operational constraints, it is used within the transfer orbit optimisation analysis. This varies the transfer orbit's apogee height, inclination and argument of perigee with a view to maximising the spacecraft mass on-station. The result consists of the optimum transfer orbit and the intermediate orbits that arise after each successive burn of the multi-burn strategy. This information is then used as initial conditions for the thrust vector steering optimisation. As the individual burns are much smaller than the single burn, it has been found that parametric apogee height reduction is unnecessary. Transfer orbit optimisation using these techniques has been carried out for ESA's OLYMPUS spacecraft.

3.4 Ground Station Coverage

The transfer orbit phase is the most demanding one in terms of ground station coverage. Many activities must be performed in a relatively short time, with the spacecraft moving rapidly through the sky. The principal activities that must be performed are:

- tracking and orbit determination,
- spacecraft telemetry monitoring and check-out,
- thruster and sensor calibration,
- appendage deployment,
- spin-up or spin-down manoeuvres,
- attitude reorientation, or slew, manoeuvres,
- orbit adjust manoeuvres,
- apogee manoeuvre(s).

Let us consider each in turn.

3.4.1 Tracking and Orbit Determination

Information on the transfer orbit characteristics is required early in this phase, and throughout it, to support activities such as slew manoeuvres, orbit adjust manoeuvres, apogee manoeuvre(s), and ground station network control. The perigee height must be known to fair accuracy (see Section 3.5) to assess whether perigee raising is necessary. With *STS* launch this is less likely than for launch by an expendable launch vehicle, because the parking orbit has already been established. Attitude reconstitution requires knowledge of the spacecraft's position with respect to the Earth and the Sun, information which is provided by orbit determination. Manoeuvre design requires a good knowledge of the current orbit, so the spacecraft attitude may be adjusted to achieve the required target orbit. For the apogee manoeuvre, when a solid propellant motor is used, only the direction of thrust may be adjusted; when a liquid propellant motor is used, both the attitude and the firing duration may be adjusted.

Orbit information also allows the control centre to calculate the direction in which ground station antennae should point to allow rapid acquisition of the spacecraft as it rises above the horizon. This information is passed on to the ground stations to allow them to operate efficiently and effectively.

3.4.2 Telemetry Monitoring

Whenever a ground station is in contact with the spacecraft, it receives its telemetry, and constantly monitors it to check the correct functioning of the spacecraft's systems. This is a vital function and allows prompt reaction to malfunctions and anomalous operation. The monitoring is done automatically, by computers examining the telemetry as it is received, checking that the parameters are within acceptable limits, and flagging and identifying discrepancies when they are detected.

Telemetry also enables the control centre to determine the spacecraft's attitude, through the output of its attitude sensors.

3.4.3 Thruster and Sensor Calibration

To ensure that manoeuvres are performed accurately, the characteristics of the thrusters used must be known accurately. This is best achieved by calibrating the thruster on orbit, in the conditions in which it will be called upon to operate. Calibration is achieved by commanding the thruster to execute a manoeuvre which is observable on the ground. Comparison of thruster action time with known initial and final states allows its delivered total impulse to be determined. This information is then used in the design of other manoeuvres performed by that thruster.

Sensor calibration serves a similar purpose. If gyros are used in transfer orbit for the control of a three-axis stabilised spacecraft, initialisation, drift-rate estimation, and scale factor estimation are all performed in orbit, with the necessary support from the ground network.

3.4.4 Appendage Deployment

This activity consists of the deployment of spacecraft appendages such as solar arrays, antennae, etc. Sometimes appendages are deployed by automatic sequences independent of the ground, while at other times deployment is initiated from the ground. Even when automatic sequences are employed, ground-controlled sequences are used as a back-up, should the automatic sequences fail.

The deployment activity consists generally of the ground station assessing that the conditions are acceptable for deployment to be initiated, if so, commanding the deployment, and verifying that it was successfully achieved.

3.4.5 Spin-Up/Spin-Down Manoeuvres

These manoeuvres are performed when the spin state at separation from the launch vehicle is different from that required during transfer orbit. If the required spin rate change is large, the main manoeuvre may be preceded by a calibration manoeuvre. The activity consists of determining the spacecraft's spin rate through examination of its sensor output. The manoeuvre to achieve the required spin rate is determined on the ground and appropriate commands are sent to the spacecraft. After the manoeuvre, the ground station determines the new spin rate. If it is within acceptable limits the manoeuvre is complete; if not, further manoeuvres are executed until the desired spin rate is obtained.

For example, the first ECS spacecraft was injected into transfer orbit by the *Ariane 1* vehicle with a spin rate of 10 rpm. For accurate firing of its solid propellant apogee motor, a spin rate of 65.5 rpm was required. The spin-up manoeuvre was executed in two phases. The first increased the spin rate to 60 rpm, while the second increased it to 65.5 rpm. The first manoeuvre was executed soon after injection into transfer orbit, while the second was carried out after first apogee passage, and after the execution of the first attitude slew manoeuvre. Such a strategy was selected to minimise the attitude de-pointing due to the spin-up manoeuvre. Also, the first manoeuvre served as a calibration for the second.

3.4.6 Reorientation, or Slew, Manoeuvres

Slew manoeuvres are required because the spacecraft attitude at injection is not that which is required for the apogee manoeuvre. The magnitude of the slew manoeuvre depends upon the launch vehicle used. The *Ariane* vehicle provides the facility to inject the spacecraft into transfer orbit at any desired attitude. This is achieved by the third stage executing a slew manoeuvre prior to separation from the spacecraft. Errors in the procedure could, however, cause an attitude error of about 6° (at the 3σ probability level). The slew manoeuvre subsequently required of the spacecraft is, therefore, only a small one. Both the *Delta* and *STS-PAM* vehicles leave the spacecraft pointing in the direction corresponding to upper stage burn-out. A large attitude change is therefore required for the spacecraft to be pointing correctly for the apogee manoeuvre. A change in spin axis direction of about 140° is required.

The ground station executes the slew manoeuvre as follows. The spacecraft's attitude is first determined by computation on the spacecraft's sensor telemetry output. Based upon knowledge of the present attitude and the desired one, the manoeuvre details are determined, care being taken that constraints such as those on the solar aspect angle are not violated during and after the manoeuvre. The manoeuvre details are then sent to the spacecraft, and the manoeuvre is commanded. As the manoeuvre is being executed, sensor telemetry is monitored on the ground to ensure that the manoeuvre is proceeding as planned. If this is not the case, the manoeuvre may be terminated upon ground command. Once the manoeuvre is complete, the spacecraft's attitude is once again computed to verify that it is as required.

The spacecraft's attitude change may be performed in several stages if the total attitude change is large. This is because, generally, the change of attitude is quite slow, and it is not possible to maintain sensor coverage of the Earth for a long time. Occasionally, slew manoeuvres, in full or in part, may be performed *blind*, i.e. outside periods of ground coverage. In such cases monitoring is not possible, but the final attitude is always checked on the ground before go-ahead for the apogee manoeuvre is given.

It is so important that the slew manoeuvres be performed accurately, that time is reserved for fine trim manoeuvres. The first slew manoeuvre is used as a calibration for subsequent manoeuvres, and hence, it is sometimes covered by two ground stations to ensure its success.

3.4.7 Orbit Adjust Manoeuvres

Two possible orbit adjust manoeuvres can be for perigee raising and apogee raising. The former is employed when the transfer orbit attained by the launch vehicle leaves the perigee too low. The consequences of this can be two-fold. The apogee is rapidly lowered due to the effect of aerodynamic drag, and the spacecraft suffers abnormally high heat input owing to friction with the upper layers of the atmosphere. To avoid both, the perigee must be raised before the spacecraft passes through it again. The most efficient point in the orbit at which to perform the perigee raising manoeuvre is at apogee. Half the period of the transfer orbit (about 5¼ hours) is, therefore, available to identify the need for such a manoeuvre, plan the manoeuvre, achieve the correct spacecraft attitude to perform the manoeuvre, and then to command its execution. This places severe demands upon the ability of the ground network. Coverage of most of the ascent to apogee is essential.

The apogee raising manoeuvre is a pre-planned part of the manoeuvre sequence, and is employed to augment the capability of the launch vehicle. The manoeuvre is performed at perigee, and since at this point the spacecraft is low and travelling very fast, ground station coverage may not be available. The manoeuvre must, therefore, either be executed *blind*, or offset from perigee to coincide with ground station coverage. The latter degrades the efficiency of the manoeuvre, but small offsets (up to about 20° in true anomaly) can usually be tolerated, as the efficiency degradation is small. Attitude manoeuvres may be performed around apogee. To support the apogee-raising manoeuvre, the ground network requires good coverage of the first transfer orbit, specially around apogee, when sensor coverage is also available.

3.4.8 Apogee Manoeuvre(s)

This is the most important activity in transfer orbit. The manoeuvre takes place near apogee, and is initiated on ground command when the controllers are satisfied that all the criteria required for its successful execution are met. The event is usually covered by two stations of the network to ensure that the execution and monitoring of the manoeuvre is not jeopardised by ground station failure. Good coverage is required up to two hours before the manoeuvre, so that the spacecraft attitude may be checked prior to the manoeuvre.

3.4.9 Coverage Limitations

The coverage a ground station can provide for a spacecraft depends upon the following factors:

- the characteristics of the orbit;
- the topography of the ground station surroundings;
- the geographical position of the ground station;

- ground station equipment limitations;
- spacecraft's telemetry, tracking and command (TTC) antenna pattern.

A geostationary transfer orbit is a highly elliptic one, with one apse at synchronous altitude. Visibility, therefore, is much greater in the region of apogee than in the region of perigee. Since the orbit apogee is very close to the Equatorial plane, and since the inclination of the orbit is small (7° for *Ariane*, 28.5^{o} for *STS*), stations near the Equator are better placed to observe the orbit than ones at high latitudes. Physical obstructions to the line-of-sight to the spacecraft, such as mountains and high buildings, effectively blot out coverage. The TTC link between the spacecraft and the ground station suffers degradation as it traverses the Earth's atmosphere. Low elevation angles are, therefore, not desirable, Most ground stations work to a minimum elevation angle of 5° for telemetry reception and command transmission. For accurate tracking, a minimum elevation angle of 10° is usually required. Finally, contact with the spacecraft is only possible through its TTC antenna. The radiation pattern of this antenna, therefore, plays a major part in determining the coverage. In transfer orbit, the spacecraft's TTC antenna should, ideally, be omnidirectional, thereby imposing no restriction on the coverage. In practice, however, this is not always possible, and cardioid patterns are common. More complex patterns with narrow but deep nulls have also been used. The assessment of ground station coverage must, in these cases, take account of the antenna pattern.

3.4.10 Coverage Analysis

The analysis of ground station coverage is fundamental to mission design, as it determines the mission sequence. The analysis is carried out by propagating the transfer orbit by small time or angle increments, and evaluating at each step whether coverage from the stations of the network is possible. The orbit propagation need not be a complex one, including only the perturbation due to the Earth's oblateness. At each step a number of important angles are evaluated. These are:

- ground station–to–spacecraft elevation angle – if this is below a specified value coverage is not obtained;
- angle between the ground station–to–spacecraft vector, and the spacecraft spin axis – to assess whether contact is allowed by the spacecraft's TTC antenna;
- angles between the spacecraft spin axis and the spacecraft–to–Earth's Limb vectors – to assess whether coverage is obtained from the spacecraft's Earth sensor;
- angle between the spacecraft spin axis and the spacecraft–to–Sun vector (also known as the Sun aspect angle) – to check that the limits placed on this parameter are not violated.

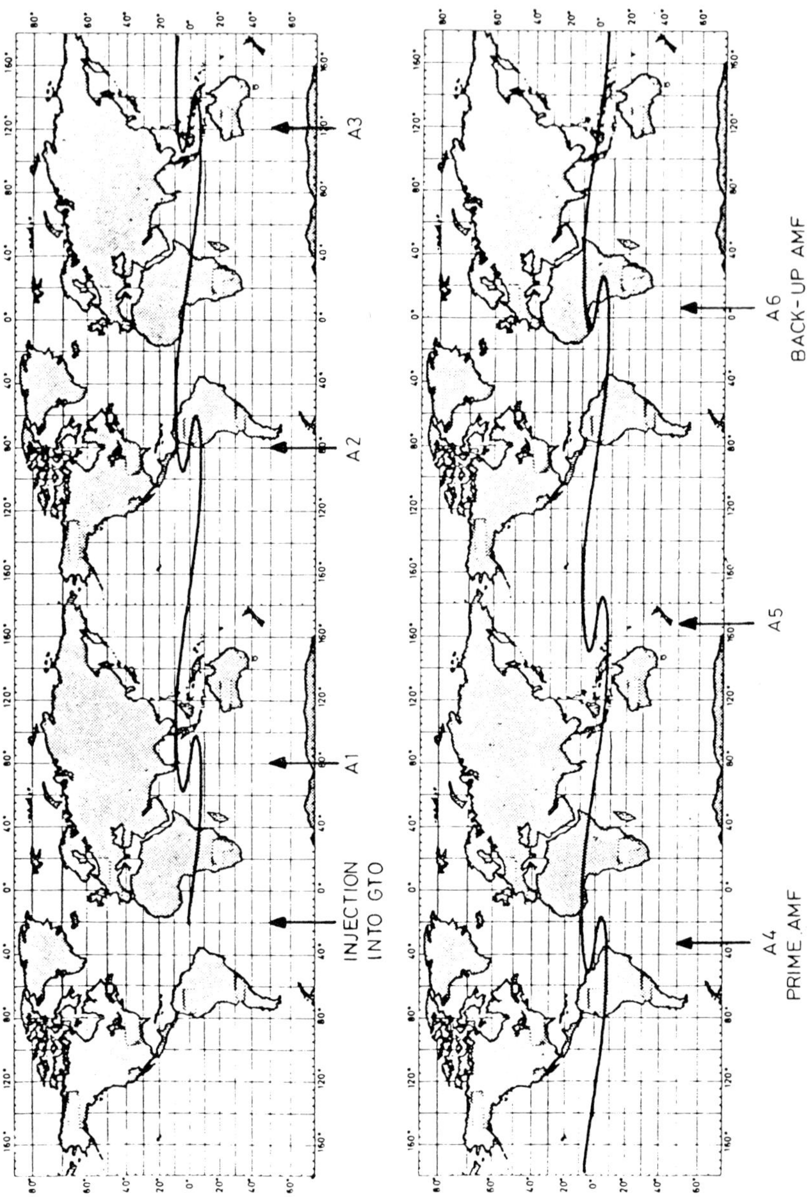

Fig.3.4 ECS Transfer Orbit Ground Track

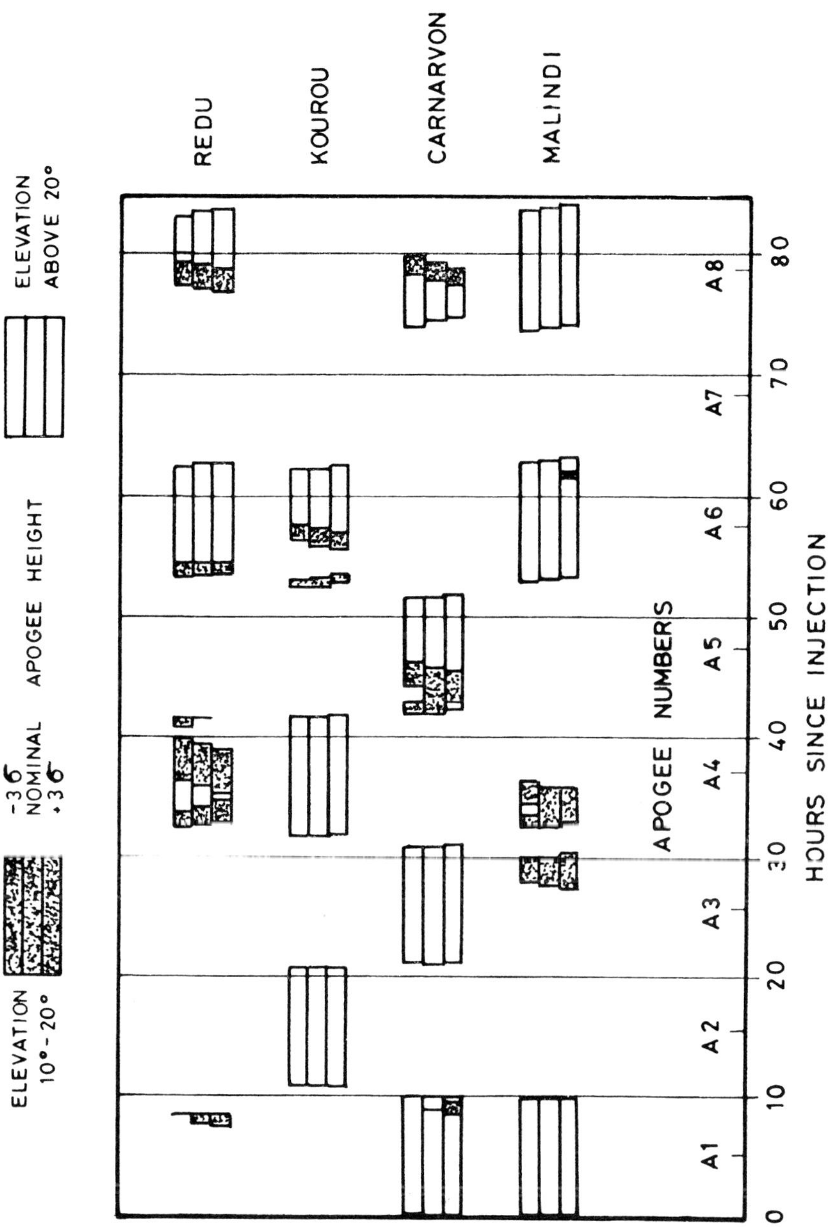

Fig.3.5 ECS Ground Station Coverage in Transfer Orbit

Other important parameters that are evaluated are:

- slant range from the ground station to the spacecraft – for TTC link budget calculations;
- the rate of change of slant range – to assess the affect of Doppler shift on the TTC link budget.

The coverage must be obtained not only for the nominal transfer orbit, but also for the most important "dispersed" cases. The orbit dispersion having the greatest effect upon coverage is the apogee height. Variation of this parameter alters the period of the orbit, and can, over a period of a few orbits, have a significant impact upon the coverage. The effects of the variation in inclination and the argument of periogee are negligible and may be ignored for the coverage analysis.

The coverage obtained for the ECS spacecraft following the ground tracks of Figure 3.4 is shown in Figure 3.5. The ground network is the ESA network comprising four stations. It can be seen that near-continuous coverage is obtained. The first slew manoeuvre occurs about 45 min. before first apogee, and is covered by two stations. The apogee manoeuvre occurs at fourth apogee, which is also covered by two stations. If this opportunity is missed, the apogee manoeuvre may be performed at sixth apogee, also covered by two stations.

3.5 Orbit Determination Requirements

The main objective of the operations during the transfer orbit phase is to achieve as good an apogee manoeuvre as possible. It is not surprising, therefore, that it is this event that places the most stringent requirements on the accuracy to be achieved by the tracking and orbit determination process. There are also, however, other requirements to be met.

Soon after the spacecraft is injected into transfer orbit it is necessary to have an estimate of the orbit elements. It is necessary to confirm at a very early stage that the transfer orbit perigee is at a height compatible with the health and well-being of the spacecraft. If the perigee is too low, the spacecraft will suffer unacceptably high heating rates due to friction which the Earth's atmosphere during perigee passage. Further, the orbit's apogee height will be reduced due to the effect of aerodynamic drag removing orbital energy during perigee passage. Knowledge of the orbit's perigee height soon after injection allows a perigee raising manoeuvre to be planned, and executed when the spacecraft is at apogee, five and a quarter hours after injection into transfer orbit. To enable the manoeuvre to be planned and executed in a timely fashion requires that the decision to implement it be made in an hour or two after injection. The ground network, therefore, has about two hours or less to track the spacecraft and determine the orbit elements. Great accuracy is not required, and it suffices to know the perigee height to within about ±5 km.

Perigee raising manoeuvres are more likely to be required with expendable launch vehicles than with the *Shuttle* for two reasons. Firstly, the *Shuttle* has already established the parking orbit, which essentially determines perigee height. Secondly, the height of the Parking Orbit is significantly higher than the perigee height achieved by expendable launch vehicles (296 km compared to 200 km for *Ariane*). In any event, the perigee raising manoeuvre is extremely unlikly to be necessary, and none has so far been implemented to the author's knowledge.

The next requirement is to determine the orbit accurately enough to allow adequate attitude determination accuracy. The accuracy requirement depends upon the sensor system and computational algorithm employed for attitude determination. A common requirement is to be able to localise the spacecraft within an error sphere of specified magnitude. The need for accurate attitude determination arises from the necessity of performing slew manoeuvres before the spacecraft attains apogee motor firing attitude. For the MARECS spacecraft, the orbit determination requirement is to localise the spacecraft within a sphere of 75 km radius with a probability of 99.7%.

Finally, we come to the most stringent requirement of all. To align the spacecraft for apogee motor firing requires a knowledge of the orbit characteristics at the time at which the motor is fired. With knowledge of the transfer orbit, the firing direction of the apogee boost motor is optimised to maximise the spacecraft's BOL mass. Once the optimum attitude has been determined attitude trim manoeuvres are implemented to bring the spacecraft to within a small tolerance of the required attitude. The current orbit is also required to be known to good accuracy to determine the firing time of the motor. Any error in knowledge of the position of the spacecraft in the orbit leads to an error in the firing time of the motor.

Below we shall outline the tracking accuracy requirements in transfer orbit of ESA's MARECS spacecraft. At the time of perigee passage preceding the apogee manoeuvre, the following tracking accuracy standard deviations were required:

- Apogee radius error $\leqslant$ 4 km
- Apogee position error $\leqslant$ 25 km
- Inclination error $\leqslant 0.04^{\circ}$
- Manoeuvre Timing error $\leqslant$ 14 sec (equivalent to an along-track error of $\leqslant 0.03^{\circ}$)

During transfer orbit, the ground network supporting the MARECS mission employed the VHF band for telemetry command and tracking purposes. Ranging accuracy at VHF is relatively poor with substantial errors arising due to propagation effects in the Earth's atmosphere. To satisfy tracking accuracy requirements, it was found that tracking was required during most of the duration of the transfer orbit.

Furthermore, to ensure that the apogee manoeuvre reliability was adequately high, it was a requirement for two stations to provide simultaneous cover before, during, and after the apogee manoeuvre, for both the prime and back-up manoeuvre opportunities.

Most modern communications satellites use S-band, C-band or Ku-band frequencies for telemetry, command, and tracking during transfer orbit. Owing to the much higher ranging accuracy possible at these frequencies, the tracking requirements can be met much more easily, and without a very strenuous programme of observations.

CHAPTER 4

THE APOGEE MANOEUVRE

4.1 Introduction

The function of the apogee manoeuvre is to place the satellite into a near-geosynchronous orbit by imparting the appropriate velocity increment at the apogee of the transfer orbit. Most transfer orbit activities are designed to achieve the prime aim of obtaining a satisfactory apogee motor firing (AMF).

The motors used for the apogee manoeuvre can be either solid propellant motors, known as apogee boost motors (ABM) or apogee kick motors (AKM), or liquid propellant motors known as apogee engines. The older generation of spacecraft favoured use of solid motors with very high thrust levels (typically 30000 N) and very short action times (typically 60 s). The velocity increment (ΔV) imparted by such a motor can be regarded as impulsive. This in turn implies that the entire ΔV can be applied at an optimum point in the orbit, and consequently the manoeuvre efficiency from the orbital mechanics viewpoint can be assumed to be 1.0 (or 100%). In addition, the orbit calculations are much simplified. The propulsion efficiency of solid propellant motors is not very high, however, and a specific impulse of around 285 s is typical. To overcome this drawback, the new generation of spacecraft has turned to the liquid bi-propellant rocket engine. The propulsion efficiency of such an engine is typically 310 s. This results in a substantial fuel saving, but owing to their low thrust (typically 490 N), the burn duration is long (typically 1 hour), and measures must be taken to ensure a high manoeuvre efficiency.

In the discussions below we shall treat solid propellant and liquid propellant apogee motors separately.

4.2 Solid Propellant Apogee Motors

Once the satellite and mission characteristics are broadly known, the search for a suitable ABM can begin. The list of suppliers is rather limited, even though the range of available motors is quite large. A shortlist of manufacturers currently supplying ABMs is:

Thiokol (US)

Aerojet (US)

Societe Europeenne de Propulsion (France)

Snia-Viscosa (Italy)

Motors exist for a whole range of satellite masses and can generally be bought off-the-shelf, though delivery times are usually several years. In certain circumstances special-to-purpose designs must be resorted to, but this is extremely expensive, as the customer pays for the test and qualification costs. Once an appropriate motor has been selected, the required propellant loading is carefully evaluated, based on the satellite mass, the required velocity increment, and the motor propellant characteristics. The exact propellant loading required is achieved by either removing some propellant (usually by machining out the core), or on-loading a larger than nominal amount. The on-loading capability of a motor design is generally smaller than its off-loading capability.

In the following sections we shall discuss the strategy adopted for selecting the burn parameters (the AMF strategy), and the assessment of satellite dynamics during AMF.

4.2.1 Apogee Motor Firing Strategy

It is the function of the ABM to achieve a near-geostationary orbit by increasing the semi-major axis of the transfer orbit to the value appropriate to a synchronous orbit, and by reducing the inclination from the transfer orbit value to a "target" value. The objective of the ABM will depend upon both the station-acquisition and station-keeping strategies (discussed in Section 5.2 and Chapter 6 respectively). For example, if full North-South station-keeping within tight limits is required during the satellite's operational phase, then the target orbit inclination will be nominally zero degrees. If a free-drift station-keeping strategy is required, the AMF target will be to place the orbit pole within the specified inclination and right-ascension limits. In addition, it may be desirable to set the AMF such that subsequent station-acquisition manoeuvres can be performed with a minimum of fuel.

The AMF strategy, and the evaluation of the desired velocity increment is closely tied with the selection of the transfer orbit, and an optimisation is usually carried out to achieve maximum satellite mass on station by varying transfer orbit parameters and the ABM velocity increment. The optimisation includes both the prime AMF opportunity and the back-up AMF opportunity (see Chapter 3).

The transfer orbit apogee number at which the ABM is fired is selected upon several considerations, and, to allow for mishaps, more than one manoeuvre opportunity is provided for. The important considerations influencing the selection of AMF apogee are:

- the apogee must be relatively close to the satellite's on-station position in order to minimise both the time and propellant requirement for station acquisition;
- there must be enough time to perform the spin axis slew manoeuvres to achieve the attitude for AMF to sufficient accuracy;
- there must be good coverage of the satellite from the ground-station network during AMF; often coverage from two stations is desirable to allow for ground-station failures;
- there must be enough time to perform the tracking operations with sufficient accuracy;
- the dwell time in transfer orbit must not be too long, as repeated traversing of the Van Allen radiation belts leads to degradation in solar array performance.

For the ECS/MARECS family of spacecraft, whose on-station longitudes are in the neighbourhood of the Greenwich meridian, the prime AMF apogee is Apogee 4, while the back-up apogee is Apogee 6. If the AMF at prime apogee is missed, a slew manoeuvre is required to align the ABM correctly for firing at the back-up apogee. This is to allow for the effect of perturbations upon the transfer orbit between the prime and back-up apogees.

The AMF strategy employed in the optimisation analysis carried out for ECS and MARECS is as follows. The position at which AMF occurs is selected to be the intersection of the transfer orbit and the *target* drift orbit. This implies that the AMF need not necessarily occur exactly at apogee. The out-of-plane component of the ABM velocity increment is chosen to achieve the target drift orbit inclination. For full North-South station-keeping missions, the target inclination is 0° while for North-South free drift missions, the inclination is the maximum allowable, e.g. 3° for MARECS. The in-plane component is therefore determined because the total velocity increment is known – i.e. it is an optimisation parameter. It remains for us to determine the spin axis elevation angle. This is determined by the requirement to minimise the fuel required for station acquisition. The station acquisition fuel requirement is calculated on the basis of a cost function dependent upon the nominal drift rate required to reach station within the stipulated time, and the apse heights of the post-AMF drift orbit.

4.2.2 Apogee Motor Firing

In order to ensure the satisfactory firing of the ABM, satellite dynamics during AMF must be investigated. Owing to misalignments, both in the thrust axis of the motor, and in the mounting of the motor relative to the satellite spin axis and centre of gravity, the satellite experiences disturbance torques about its principal axes. As the satellite is spin-stabilised during AMF, these disturbance torques lead to *nutation* of the satellite's spin axis. This in turn leads to the thrust being applied in a non-nominal direction. This is equivalent to a loss of thrust in the required direction, which must be accounted for when specifying the velocity increment requirements of the ABM.

There are, in fact, two reasons why the ABM thrust is not applied in the required direction. One is explained above, while the other is the uncertainty in the knowledge of the spin axis direction. The thrust direction error due to the two effects acting concurrently must lie within a specified limit. For ECS, for example, this limit was 1° at the 99.7% probability level. This amount must be apportioned between the two effects such that:

$$\Delta\theta^2 = \Delta\theta_1^2 + \Delta\theta_2^2$$

where: $\Delta\theta$ = total thrust vector direction error (at 3σ level);

$\Delta\theta_1$ = thrust direction error due to attitude reconstitution error (at 3σ level);

$\Delta\theta_2$ = thrust direction error due to thrust misalignment etc. (at 3σ level).

Attitude reconstitution error analysis will lead to a definition of $\Delta\theta_1$, and this will in turn define $\Delta\theta_2$ when the overall accuracy requirements are known. In order to assess whether the $\Delta\theta_2$ requirement can be met, a simulation of the satellite dynamics during AMF is carried out using a realistic model of the thrust vector alignment errors. Below we shall give a brief outline of the theory employed.

The Euler equations of motion for the satellite are:

$$\begin{bmatrix} I_1\dot{\omega}_1 \\ I_2\dot{\omega}_2 \\ I_3\dot{\omega}_3 \end{bmatrix} = \begin{bmatrix} 0 & \omega_3 & -\omega_2 \\ -\omega_3 & 0 & \omega_1 \\ \omega_2 & -\omega_1 & 0 \end{bmatrix} \cdot \begin{bmatrix} I_1\omega_1 \\ I_2\omega_2 \\ I_3\omega_3 \end{bmatrix} + \begin{bmatrix} N_1 \\ N_2 \\ N_3 \end{bmatrix} - \begin{bmatrix} (\dot{I}_1 - M(d^2 + \frac{1}{2}r^2))\omega_1 \\ (\dot{I}_2 - M(d^2 + \frac{1}{2}r^2))\omega_2 \\ (\dot{I}_3 - Mr^2)\omega_3 \end{bmatrix} \quad (1)$$

where: ω_1, ω_2 and ω_3 are the spacecraft angular rates about the X, Y and Z axes – the Z-axis being the spin axis of the body axis set;

I_1, I_2 and I_3 are the principal moments of inertia of the spacecraft;

d is the distance between the body's centre of gravity and the ABM nozzle exit plane;

r is the mean effective radius of the ABM nozzle;

N_1, N_2 and N_3 are the disturbance torques due to the thrust vector misalignment;

M is the mass flow rate from the ABM nozzle (a negative quantity).

The vector (N_1, N_2, N_3) is obtained from a knowledge of the thrust vector misalignment from the nominal and the displacement of the actual centre of gravity from the nominal position. Several approaches to the evaluation are possible. As the misalignments and errors are of a random nature it is possible to employ a Monte-Carlo procedure to calculate the disturbance torque components constituted of all the elements of the misalignments. In such a procedure it is recognised that certain variables conform to normal distributions, while others conform to rectangular distributions. With this method, the distributions of the disturbance torque components can be evaluated, and from them the 99.7% probability values can be obtained. Such a method is, however, lengthy and expensive on computer time. To overcome this drawback, a simpler method may be used. This consists of evaluating the 99.7% probable value of the disturbance torque components, based upon the 99.7% probable values of all the individual errors and misalignments that contribute to them. We make the assumption that all the errors are uncorrelated, and combine them by a simple *root-sum-square* (RSS) procedure. In order to be able to use the simpler method with confidence, its results must be checked against the more thorough Monte-Carlo method. This was done for OTS, and it showed that for the resultant moment arm, at the 99.7% probability level, the two methods agreed to within 6%. The RSS method was conservative, in that it predicted a higher moment arm, resulting in a greater disturbance torque. The RSS method can, therefore, be used to evaluate the vector (N_1, N_2, N_3).

The errors that are considered in the evaluation of the disturbance torque vector are:

- the thrust vector lateral misalignment relative to the ABM case (l_1);
- the lateral misalignment of the ABM case relative to the spacecraft nominal spin axis (l_2);
- the lateral misalignment of the spacecraft centre of gravity (l_3).

Using the RSS method, the above errors are combined to give the total lateral misalignment, $l = (l_1^2 + l_2^2 + l_3^2)^{1/2}$, and the direction of l in the X–Y plane is defined by the angle α, measured from the satellite +X axis. See Figure 4.1.

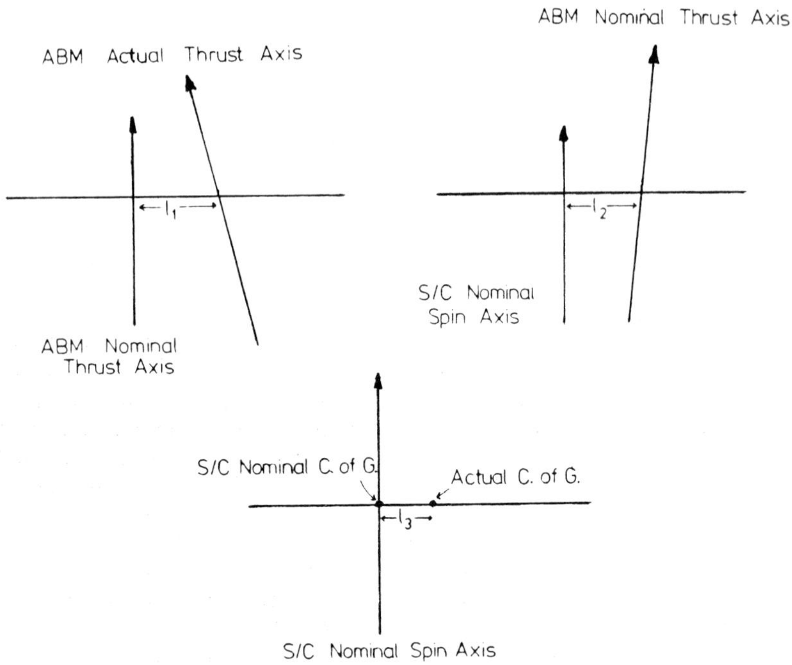

Fig.4.1 Position Misalignments of Thrust Vector

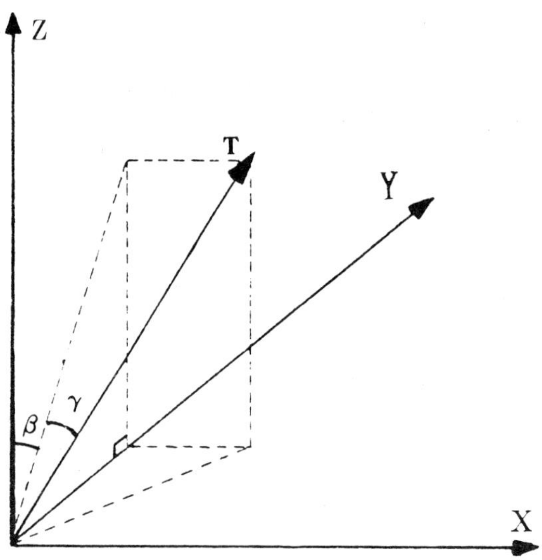

Fig.4.2 Angular Misalignments of Thrust Vector

The angular errors are defined by Figure 4.2 as follows:

- the angular misalignment of the thrust vector relative to the spacecraft X–Z plane (β);
- the angular misalignment of the thrust vector relative to the spacecraft Y–Z plane (γ).

β and γ define the RSS combination of the angular misalignment of the thrust vector relative to the ABM nominal thrust axis and the angular misalignment of the ABM nominal axis relative to the spacecraft nominal spin axis.

Using the 3σ values for all the error components, the disturbance torque vector is evaluated to be:

$$\begin{bmatrix} N_1 \\ N_2 \\ N_3 \end{bmatrix} = T \begin{bmatrix} l \sin a \cos \beta \cos \gamma + h \sin \beta \cos \gamma \\ -h \sin \gamma - l \cos a \cos \beta \cos \gamma \\ l \cos a \sin \beta \cos \gamma - l \sin a \sin \gamma \end{bmatrix} \qquad (2)$$

where: T = thrust of the motor

h = the distance between the point of application of the thrust and the spacecraft centre of gravity measured along the Z axis.

Eqn. (2) is employed in the Euler equations – Eqn. (1).

The final terms of Eqn. (1), i.e. $M\,(d^2 + \frac{1}{2}r^2)\,\omega_1$, etc. are known as the *jet damping* terms, and arise from the angular momentum transferred to the motor exhaust plume. If the spacecraft ejects mass from the motor nozzle at a rate M, then the rate of change of angular momentum of the exhaust plume, about each ot the spacecraft axes, is equal to $-Md_i^2\,\omega_i$, where d_i is the mean effective distance of the exhaust plume from the axis and ω_i is the spacecraft angular rate about the same axis. This is equivalent to a torque of the same magnitude acting on the spacecraft. As this effect reduces the oscillation of the transverse angular velocities of the satellite, it is referred to as a *damping* effect.

The Euler angles θ, ψ, and ϕ relate the spacecraft's body axes to a set of axes fixed in inertial space. The Euler equations of motion can be integrated numerically to yield the motion of the spacecraft. It can be shown that:

$$\dot{\theta} = \omega_2 \cos \varphi - \omega_3 \sin \varphi$$

$$\dot{\psi} = (\omega_2 \sin \varphi + \omega_3 \cos \varphi)/\cos \theta$$

$$\dot{\varphi} = \omega_1 + \dot{\psi} \sin \theta$$

The vector representing the change in velocity produced by the ABM is given by:

$$\Delta V = \int_{o}^{t_b} T'(t) \quad dt = (\Delta V_1, \Delta V_2, \Delta V_3)$$

where: t_b = burn duration

$T'(t)$= the thrust vector in satellite body axes

$M(t)$ = mass of spacecraft

The mean direction of the spin axis is at an angle θ^* to the nominal direction, where:

$$\tan \theta^* = \sqrt{\Delta V_1^2 + \Delta V_2^2} / \Delta V_3$$

The nutation angle, η, is defined to be:

$$\tan \eta = (I_1^2 \omega_1^2 + I_2^2 \omega_2^2)^{1/2} / I_3 \omega_3$$

To increase the accuracy of the simulation, a representative thrust profile should be used, and provision must be made to include non-zero initial angular displacements and rates.

A simulation such as the above will yield the following useful information:

- end-of-burn velocity increment components in satellite body axes;
- end-of-burn velocity increment components in inertial axes;
- the nutation angle history;
- the history of the velocity error angle, i.e. θ^*;
- the dynamic behaviour of the satellite during the burn.

The components of velocity increment in inertial axes can now be compared with the ABM velocity increment for a perfectly aligned ABM. They will show some differences from the nominal values. This indicates the worst case performance. The designer may choose to counteract this loss of performance by on-loading the ABM by the corresponding amount, or he may choose to meet the requirements by increasing the satellite spin rate. For OTS, the ABM was on-loaded to counteract the effect of the nutational losses.

4.3 The Liquid Propellant Apogee Engine

The previous discussion within this chapter relates entirely to solid propellant apogee motors. These can now be regarded as *classical*, and many new designs are now based upon the use of bi-propellant technology for the apogee boost function. The main advantages of the bi-propellant apogee engine are the following:

- lower thrust level (important for the future concept of highly flexible spacecraft);
- higher propulsive efficiency;
- higher injection accuracy;
- multiple burns possible, through motor re-start;
- *parasitic* station acquisition possible (see Section 4.3.1.2);
- common system with attitude and orbit control functions.

The main disadvantages of the system are:

- increased system complexity;
- increased sub-system cost;
- somewhat greater operational complexity;
- slightly lower manoeuvre efficiency owing to the longer duration of the burn(s).

Selection of the system depends upon the overall system-level trade-off, and several spacecraft have selected the bi-propellant system, e.g. INSAT, OLYMPUS, INMARSAT, and INTELSAT 6.

The possibility of executing the apogee manoeuvre in more than one burn, and the long burn duration, have opened two new areas of analysis:

- selection of the number of burns, partitioning of the complete manoeuvre between the burns, and the timing of the burns;
- optimisation of thrust vector steering to obtain the maximum efficiency.

These analyses are too complex to be done manually. Furthermore, they need to be carried out more than once during the life of a spacecraft project to keep step with changes in design and requirements. It is essential, therefore, that suitable software be developed to make the mission design task accurate and efficient. It has been found useful to have software available to carry out the following evaluations:

- analysis of overall multi-burn strategy incorporating *parasitic* station acquisition;
- *Ariane*-related transfer orbit optimisation using the multi-burn apogee engine firing (AEF) strategy;
- *Shuttle*-related version of the above;
- single-burn thrust vector steering optimisation incorporating several steering laws;
- multi-burn thrust vector steering optimisation using the *Pontryagin* optimisation technique.

4.3.1 Apogee Engine Firing

4.3.1.1 The Single-Burn Strategy

The mission considerations related to a single-burn AEF strategy are very similar to that for a solid propellant motor strategy. The firing time must be selected with a view to:

- good ground station coverage for attitude manoeuvres;
- good ground station coverage for orbit tracking;
- proximity to the on-station longitude;
- minimum dwell time in transfer orbit consistent with the above;
- adequate opportunity for a back-up manoeuvre in case of the complete, or partial, non-execution of the prime manoeuvre.

If the transfer orbit mode is spin-stabilised, adequate account must be taken of the time and coverage required for the spin axis slew manoeuvre(s). For a 3-axis stabilised transfer orbit mode, adequate time must be allowed for sensor calibration.

Most of the analysis related to a single-burn strategy is associated with the optimisation of the steering law. For OLYMPUS, the steering law selected consists of a constant rate slew about a single axis fixed in inertial space. Such a law was selected because it was easy to implement, and also gave an adequately high manoeuvre efficiency compared to an impulsive burn, i.e. 99.5%.

The back-up cases that have to be considered must include not only the case where the prime manoeuvre failed to be initiated, but also cases of incomplete execution of the prime burn. In the case of the latter, the subsequent manoeuvre must take place in an orbit intermediate between the transfer and drift orbits, and the manoeuvre design must take into account, in the best possible way, all the constraints that played a part in the design of the prime manoeuvre itself. Such an analysis was carried out for OLYMPUS and it was found that, in general, the later in the prime manoeuvre that the abort occurs, the later the opportunity for the follow–on manoeuvre occurs. For example, for aborts 20% and 40% into the prime burn, the opportunity for the follow–on manoeuvre occurs approximately 2 orbit periods later. For aborts 60% and 80% into the prime burn, follow–on burn opportunities occur 3 and 4 orbit periods later, respectively.

In designing the follow–on manoeuvres account must be taken of the time required for diagnosis, error correction, thrust vector steering evaluation, and orbital operations.

4.3.1.2 The Multi–Burn Strategy

The multi–burn strategy poses rather more complex design problems, but its advantages are significant. The main advantages are that:

- because the arc over which each burn is made is smaller, the burn efficiency is greater;
- the engine can be calibrated during the first burn, making the subsequent burns more accurate;
- subsequent burns can be used to compensate for errors in previous burns, making orbit injection more accurate;
- by varying the magnitude and timing of the individual burns, *parasitic* station acquisition can be performed, resulting in very significant fuel savings.

However, mission complexity is undeniably increased, though not to unacceptable levels. A further adverse effect of the multi–burn strategy is that the acceleration loading on the spacecraft (due to transient effects) is significantly higher during the last burn of a multi–burn strategy than it is for a single–burn strategy. If the transfer orbit mode is 3–axis stabilised with deployed solar arrays, the extra bending load on the arrays could impose a mass penalty. Such considerations play their part in the overall, system level, configuration trade–offs. The factors affecting selection of the details of the multi–burn strategy are outlined below.

The Number of Burns

If two burns are used, each would be rather large, and manoeuvre efficiency would suffer. The errors in the second burn would be large leading to a large station acquisition fuel requirement.

If three burns are selected, each burn is much smaller than a single burn and hence more efficient. Further, errors in the first burn can be removed by the second and errors in the second by the third, thus achieving a progressive error removal. Because the final burn is small, its errors are small too, and hence the station acquisition fuel requirement is small.

If four or more burns are selected, the strategy becomes operationally very inconvenient, with a high ground station work load. Motor reliability also decreases as the number of motor starts increases.

This indicates that a three-burn strategy is probably the most attractive.

The Burn Sizes

In general, the burn sizes should be approximately the same, with the last burn rather smaller than the other two. This has the advantage that all the burns have a high efficiency, and the burn errors of the last burn are small.

For OLYMPUS launched on *Ariane*, analysis of a three-burn strategy incorporating *parasitic* station acquisition led to the following burn characteristics:

	Burn 1	**Burn 2**	**Burn 3**
Velocity Increment (m/s)	504	761	249
% of total	33%	50%	17%
Executed at apogee number	2	4	5

Execution Times

The execution times are selected in a manner than allows *parasitic* station acquisition. This implies that burn sizes and coast phases are adjusted so that the final burn occurs at the desired on-station longitude. Only errors in the final burn, therefore, need be corrected by the subsequent station acquisition phase.

Other points to be considered are:

- sufficient time must be allowed before each burn for orbit tracking and manoeuvre planning;
- sufficient time must be allowed before each burn for attitude sensor calibration, slew manoeuvre execution, and manoeuvre parameters up-linking;
- sufficient ground station coverage must be available for the above activities.

Other Factors

In addition to the above, the selected strategy must be such as to provide adequate back-up strategies at any point in the execution of the prime strategy, to allow recovery from failure to initiate or to complete a burn. The back-up strategies must themselves be as fuel-efficient as possible.

4.3.1.3 Analysis Techniques

The Single-Burn Manoeuvre

The optimisation of the thrust vector history for the single-burn strategy should consider a number of alternative steering laws. These are:

- continuously variable unconstrained thrust direction;
- thrust direction resulting from constant rates about two orthogonal inertially-fixed axes;
- thrust direction resulting from a constant rate about a single inertially-fixed axis;
- inertially fixed thrust direction.

The optimisation variables are those associated with the steering laws, and the position in the orbit at which the burn is initiated. The constraints apply to the end-of-burn orbit elements and ensure that the target orbit is achieved within specified tolerances. The target and pre-burn orbits are input data, and the optimisation aims to minimise the burn duration and hence the fuel requirement for the manoeuvre. Owing to the nature of the problem, convergence to the optimum can be quite slow, and good estimates of initial values are required.

The Multi-Burn Manoeuvre

The optimisation of the multi-burn apogee manoeuvre is somewhat more complex owing to the large number of variables involved. There are a large number of alternative multi-burn strategies that permit *parasitic* station acquisition in a fuel-efficient manner. Not all the alternatives will, however, be acceptable from an operational viewpoint. Constraints such as ground station coverage, total elapsed time of the apogee manoeuvre, and event scheduling, must be respected. The large number of alternatives are, therefore, reduced considerably, and a short-list of three or four may be taken forward to the next level of detailed analysis involving the optimisation of the thrust vector pointing history for each of the burns. The final choice is then made on the basis of maximum manoeuvre efficiency.

We shall now outline the manner in which this two-stage procedure is carried out. The first stage consists of identifying the large number of multi-burn strategies that permit *parasitic* station acquisition in a fuel-efficient manner. To enable this to be done rapidly and efficiently, a number of simplifying assumptions are made. These are:

- the burns are considered to be impulsive;
- the burns occur at the node of the transfer orbit nearest to apogee;

- the size of the first burn is kept constant (but is selectable);
- the final two burns are varied in size to achieve *parasitic* station acquisition;
- the total burn magnitude and direction of the multi-burn strategy are defined by the need to transition from a defined transfer orbit to a geo-synchronous orbit;
- the J_2 (oblateness) harmonic perturbation to the orbit need not be considered in these calculations.

Once the alternative strategies are reduced to a short-list, the second stage can begin. The second stage achieves the optimisation of the thrust vector pointing history for each of the burns. The optimisation software uses the intermediate orbits derived in the previous stage as input data. The optimisation software should be capable of considering several thrust vector steering strategies. Among the most useful are:

- inertially-fixed pointing of the thrust vector during a burn – the inertial direction of each burn may be different;
- constant slew rate about a single axis fixed in inertial space;
- unconstrained thrust vector (the *Pontryagin* approach).

The final selection is based upon fuel-efficiency and operational considerations.

It should be noted, that other, more direct, techniques are also available, and may be considered. The method outlined here gives intimate control over strategy selection, and introduces operational constraints at an early stage in the analysis.

4.3.2 Operational Considerations

Here we shall consider some of the operational aspects of the liquid propellant apogee engine manoeuvre. Though the transfer orbit optimisation does take into account launcher dispersions, the manoeuvre details are evalauated for only the nominal and other selected cases. During the mission itself the achieved transfer orbit will be to some degree off-nominal. We, therefore, require some means to take account of this. Some of the questions that arise are:

- Can we apply the nominal manoeuvre to an off–nominal initial orbit with acceptable accuracy?
- Is the tracking available between burns of adequate accuracy to justify re–optimisation of the following burn?
- Do we have enough time to carry out a re–optimisation between burns and implement the revised manoeuvre?

To answer some of these questions an investigation was carried out for the EUROSTAR spacecraft. The scenario considered was of a *Shuttle/PAM–D2* launch, with transfer orbit control provided by the COMSAT network employing ground stations at Southbury (U.S.A.), Santa Paula (U.S.A.), and Guam. The apogee manoeuvre consisted of three burns of roughly equal magnitude. The following results were obtained, and are believed to be typical.

(a) For 3σ transfer orbit dispersions in conjunction with a nominal apogee manoeuvre strategy, unacceptably high station–acquisition fuel useage resulted. Manoeuvre re–optimisation is, therefore, necessary.

(b) If 3σ propulsion and attitude errors during the burns are not compensated by manoeuvre re–optimisation, considerable fuel useage occurs during station acquisition, leading to a degradation of beginning–of–life mass.

(c) When the re–optimisation was executed using a circular geosynchronous target orbit, convergence of the optimisation algorithm was difficult, the resultant thrust vector directions exhibited a large variation ($\pm 15^{\circ}$) from the nominal, and propellant useage was high.

(d) When the target orbit was changed to an elliptical one, with the height of one apse equal to that of the transfer orbit, and the height of the opposite one equal to geostationary height, convergence of the optimisation algorithm was very much more rapid. In addition, the thrust vector direction exhibited a much smaller variation ($\pm 3^{\circ}$) from the nominal, and fuel useage was lower.

Based on these results, the following operational methodology was evolved. To maximise the spacecraft's beginning–of–life mass, it is necessary to re–optimise the multi–burn apogee manoeuvre based on the actual transfer orbit derived from tracking data, and the elliptic target orbit described in (d) above. Following the execution of a burn, the above procedure is repeated until all the burns have been completed. Then station–acquisition manoeuvres are implemented to bring the spacecraft to its on–station longitude. The re–optimisation computer times are such that this procedure is a feasible one, though vigorous tracking is required. At this point it is worth noting that had the target orbit defined in (c) above been adopted, the resultant variation in thrust vector attitude would have resulted in near-total closure of the launch window.

4.4 Further Reading

The reference below contains a discussion on the nutation dynamics of the spacecraft under thrust:

(a) Thomson, W.T. : 1963, "Introduction to Space Dynamics", Published by J. Wiley and Sons Inc., New York.

The following reference considers the influence of liquid propellant motion on the dynamics of a spin-stabilised spacecraft during the burn duration of its solid apogee motor:

(b) Ono, S. : 1985, "Spindynamics with Sloshing Effects for Solid Motor Burning Phase", *Proceedings of the International Conference on Space Dynamics for Geostationary Satellites*, Toulouse, October 1985.

The reference below outlines in detail the simulation of the operational implementation of a multi-burn apogee manoeuvre and forms the basis upon which the conclusions of Section 4.3.2 are drawn:

(c) Pocha, J.J. and Weber, M. : 1985, "Operational Strategies for Multi-Burn Apogee Manoeuvres of Geostationary Spacecraft", *Proceedings of the International Conference on Space Dynamics for Geostationary Satellites*, Toulouse, October 1985.

CHAPTER 5

DRIFT ORBIT

5.1 Introduction

A successful apogee manoeuvre places the satellite in a near-geosynchronous orbit, also known as a *drift orbit* owing to the slow drift of the satellite in longitude. The main mission events during the drift orbit phase are *attitude acquisition*, *station acquisition*, and *initialisation*.

Attitude acquisition consists of a sequence of manoeuvres the spacecraft undertakes to convert its attitude stabilisation mode from that adopted for transfer orbit to that which is employed during the operational phase of the mission. The two modes of stabilisation can be, and usually are, very different. For spacecraft employing solid propellant apogee boost motors, spin stabilisation is employed to give the necessary stability for the correct firing of the motor. For spacecraft employing liquid propellant apogee engines the transfer orbit stabilisation mode may be either spin-stabilised or three-axis stabilised. The stabilisation mode for the operational phase depends upon the configuration of the spacecraft. A large family of spacecraft are spin-stabilised, employing mechanically or electronically de-spun antennae for on-station operations. Others are three-axis stabilised. The ECS/MARECS/SKYNET/TELECOM family of spacecraft is three-axis stabilised in the operational orbit. For these spacecraft, which are spin-stabilised during transfer orbit, transition from the spin-stabilised mode to the three-axis stabilised mode involves de-spin, Sun acquisition, Earth acquisition, and initiation of on-board systems for *normal mode* operations. The requirements for the various manoeuvres are specified by the attitude control subsystem. The manoeuvres must be carefully scheduled during the drift orbit phase. There is usually a time limit for the establishment of the operational attitude, and the manoeuvres must be organised to satisfy this limit.

As far as the mission analysis is concerned, the attitude acquisition activity is one of scheduling, and co-ordination with periods of sensor and ground station coverage. Once the satellite is in its operational attitude, it is able to undertake the station acquisition manoeuvres. This is the subject of the following section.

5.2 Station Acquisition

It is the objective of the station acquisition manoeuvres to place the satellite at its operational longitude within allowed limits, and to achieve a geostationary orbit, again within allowed limits. These limits are dictated by the satellite's station-keeping strategy, which is discussed in Chapter 6.

In general, the post-apogee manoeuvre drift orbit will be elliptic, and of non-zero inclination. The station acquisition phase will, therefore, consist of manoeuvres designed to achieve:

- the required orbit inclination;
- the drift rate necessary to place the satellite at the required longitude;
- geostationary orbit once the satellite is at the correct longitude.

A limit is usually imposed upon the time available for station-acquisition manoeuvres, as the satellite is required to begin its useful operations – its communications mission – as soon as possible. To achieve the above objectives, manoeuvres are required both in the orbit plane and perpendicular to it. The manoeuvres are performed with the satellite's reaction control thrusters. We are, therefore, dealing with a low-thrust system with finite burn times. However, as the required velocity increments are also quite small, it is a good approximation, for design purposes, to consider the burns to be applied impulsively at the optimum point of the orbit. Since the in-plane and out-of-plane velocity increments affect different orbit elements, we can consider them to be de-coupled. With most spacecraft, the in-plane and out-of-plane manoeuvres will be implemented separately, with separate thrusters.

5.3 Station Acquisition Error Analysis for Spacecraft with Solid Propellant Apogee Motors

For mission design purposes, the most important consideration regarding station acquisition is the amount of fuel to be allocated to achieve acquisition with a probability of, say 99%. To determine this we are obliged to proceed on a statistical basis, and take account of the dispersions that occur about the nominal trajectory, from launch to the acquisition of the on-station longitude. A Monte Carlo simulation technique is employed. In essence, this consists of simulating the same sequence many times, the difference between each simulation being the magnitudes of the various dispersions. The dispersions are assumed to conform to specified probability distributions. In each simulation, random numbers are selected in such a way that their probability of occurring is consistent with their probability distribution. The main result of each simulation is the total station acquisition velocity increment. Several simulations thus provide a large sample of velocity increments from which conclusions may be drawn. Since both analysis cost and accuracy are directly related to sample size, a compromise is necessary, giving results of acceptable accuracy at reasonable cost. For ECS and MARECS, for example, a sample of 2000 simulations gave acceptable accuracy.

The modelling of station acquisition consists of the following elements:

- transfer orbit dispersions due to injection errors;
- errors in the tracking of the transfer orbit;
- apogee motor firing strategy;
- errors in the above, resulting in dispersions in the drift orbit;
- station acquisition manoeuvre sequences yielding a minimum velocity increment to acquire station from a given set of drift orbit conditions.

Owing to the launch vehicle dispersions, the actual transfer orbit elements will be subject to variation from their nominal values. The covariance matrix of launch vehicle errors gives a statistical measure of the possible variation of the transfer orbit – see Section 2.4. Using the covariance matrix, nominal transfer orbit elements, and appropriate software, a set of random transfer orbits is generated.

By the time the satellite reaches the apogee selected for the apogee manoeuvre, the orbit will have been tracked from the ground, and will be known to within a certain accuracy. An estimate of this accuracy is made by the tracking error assessment software – see Section 9.3.3 – and will result in a covariance matrix which gives a statistical measure of the uncertainty in the derived values of the orbital parameters.

For the apogee manoeuvre, the spacecraft will have been oriented to the correct attitude. The optimum attitude is that which, after an error-free burn, results in a minimum velocity increment required for station acquisition. The spacecraft attitude is defined by two angles: either the right ascension and declination of the spin axis, or its azimuth and elevation. The apogee manoeuvre strategy employed has been discussed in Chapter 4, and will only be summarised here.

Because orbit inclination errors are expensive to correct, the required inclination of the geosynchronous orbit is aimed for. The apogee manoeuvre must, therefore, occur when the transfer orbit crosses the required geosynchronous orbit plane. It can be shown that, to a high degree of accuracy, the total in-plane station acquisition velocity increment can be expressed as an analytic function of:

- the separation between the on-station and apogee manoeuvre longitudes;
- the radii of the post-manoeuvre drift orbit apses.

The apse heights can, in turn, be related to the orientation of the spacecraft during the apogee manoeuvre. By minimising the station acquisition velocity increment as well as aiming for the required inclination of the geosynchronous orbit, the optimum alignment is explicitly determined.

The apogee manoeuvre actually implemented will, however, differ from the nominal due to the influence of several error sources. These error sources are modelled in a statistically accurate manner, and the post-manoeuvre drift orbit is obtained for each simulation. The error sources considered are enumerated below.

(1) Attitude Reconstitution Error

Knowledge of the spin axis attitude will be inaccurate owing to the errors associated with the attitude reconstitution process.

(2) Transfer Orbit Determination

The optimum alignment of the spacecraft and the timing of the apogee manoeuvre will be determined by an estimate of the transfer orbit. As this is slightly different from the actual orbit, some errors will occur.

(3) Apogee Manoeuvre Command

The command to fire the apogee motor is time–tagged, but can also be directly initiated from the ground. Some timing error will occur.

(4) Slew Manoeuvre Error

There will be some error in achieving the desired attitude owing to the quantisation of the attitude reorientation manoeuvre implementation.

(5) Motor Firing Dynamics

Motor and satelllite misalignments will cause the thrust to be delivered in a non–nominal direction – see Section 4.2.2.

(6) Apogee Motor Impulse

There is a tolerance associated with the nominal performance of the motor in terms of the velocity increment it delivers. The major contribution to the tolerance is the uncertainty in the specific impulse.

Some of these error sources, i.e. attitude errors, specific impulse errors, and timing errors, have normal distributions, while others, i.e. manoeuvre errors, have rectangular distributions. Some errors, i.e. the firing dynamics errors, follow Rayleigh distributions.

The orbit following the apogee manoeuvre will, in general, be elliptic with an inclination other than that desired. The station acquisition manoeuvres will achieve the required operational orbit. The in-plane manoeuvres will be performed at the drift orbit apses, and will consist of Hohman transfers. The out-of-plane manoeuvre will be performed at the intersection of the drift orbit and the required geosynchronous orbit. A minimum of three in-plane and one out-of-plane manoeuvres is necessary. The manoeuvres are planned with the object of minimising the expenditure of fuel without violating the constraints imposed upon burn timing and ground station visibility.

There are six possible basic combinations of the sequence of the three in-plane burns. The purpose and location of the final manoeuvre is common to all six. It is performed at the apse which is at geostationary height at the station longitude, and sets the opposite apse to geostationary height, thus circularising the orbit. The function of the other two burns is to set one apse at geostationary height, and to set up the required drift rate.

The position and order of these two burns is altered to give the six combinations. For each of the six combinations, there are several possibilities for the timing of the manoeuvres. This is because, although the position in the orbit at which a manoeuvre is made is fixed, there is freedom of choice in the number of orbits between manoeuvres. From the host of available possibilities, the one that is chosen is that which gives the minimum velocity increment for station acquisition, and satisfies all the constraints. These constraints are discussed next.

The main constraint is the requirement to arrive on station within a specified time after the apogee manoeuvre. For ECS and MARECS, this time limit was 21 days. The time limit is generally in the region of 21 days because it can be shown that this point is at or very near the "knee" of the curve between time period and station acquisition velocity increment. A investigation carried out for UNISAT gave the curve shown in Figure 5.1. Below 21 days, the station acquisition velocity increment rises substantially. Beyond 21 days, the reduction in station acquisition velocity increment is negligible. Another constraint arises from the need to track the orbit after manoeuvres, to confirm the success of the manoeuvre, and as a starting point for planning the next. A certain amount of time is, therefore, reserved for this activity and results in a constraint upon the minimum time between manoeuvres. Further, allowance must be made for the time required for attitude acquisition and initialisation. The latter comprises a series of manoeuvres, usually spanning 24 hours, designed to set up the orbit elements appropriate to the start of the station-keeping operations. The time within which the station acquisition manoeuvres must be performed is therefore somewhat reduced. Another constraint is the requirement for ground station coverage for manoeuvre initiation. This is discussed in Section 5.6. It does not however, adversely affect station acquisition fuel requirements.

The manner in which the out-of-plane manoeuvre is interleaved with the in-plane manoeuvres will depend upon the magnitude of the respective manoeuvres. If the out-of-plane manoeuvre is small compared to the in-plane manoeuvres it should be timed to take place after the first two in-plane manoeuvres. Then any out-of-plane component arising from these two manoeuvres, due to misalignment,

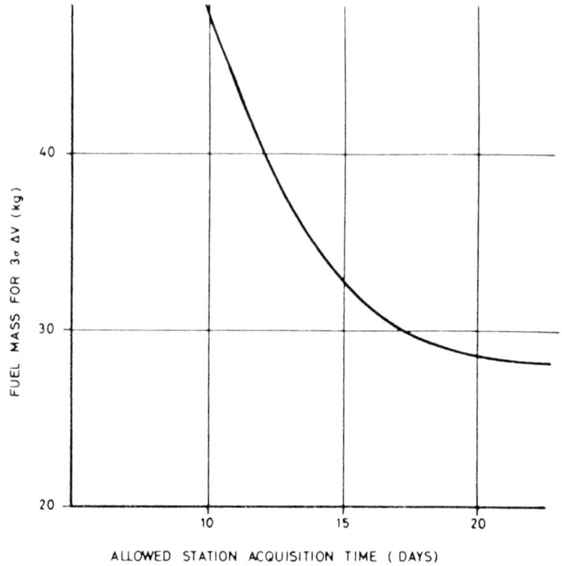

Fig.5.1 Station Acquisition Propellant Versus Acquisition Time

can be corrected. If the out-of-plane manoeuvre is greater than the in-plane manoeuvres, it should be performed as early as possible. Then any in-plane component resulting from the burn can be compensated by adjusting the in-plane manoeuvres.

The evaluation of the station acquisition velocity increment on a statistical basis is programmed for a digital computer. Below we shall outline the main steps in the procedure.

- Using the launch vehicle covariance matrix, and the nominal transfer orbit parameters, N sets of errors in the transfer orbit elements are generated.

- Using the tracking error covariance matrix, and the nominal transfer orbit parameters, N sets of errors in the estimation of the transfer orbit elements are generated.

- The information produced above is used, together with apogee motor and orbit data, in a program which implements the apogee manoeuvre according to the strategy discussed previously, and generates N sets of drift orbit parameters.

- These drift orbit parameters are used, together with station acquisition constraints and target orbit information, in a program which executes the station acquisition manoeuvres. For each drift orbit, all six manoeuvre strategies are tried, and the one that yields the lowest velocity increment is selected.

The principal output of the final step is the cummulative probability distribution of the total station acquisition velocity increment. From this, the velocity increment required to accomplish station acquisition to a certain probability level, can be obtained easily. This information is used directly to determine the propellant mass to be allocated for the station acquisition manoeuvres. Detailed information about each simulation can also be obtained in order to aid in the planning of the operations for the station acquisition phase of the mission.

5.4 Station Acquisition Error Analysis for Spacecraft with Liquid Propellant Apogee Engines

The considerations and the method of analysis described above for the station acquisition error analysis when the apogee motor is a solid propellant one, do not apply for a multi-burn apogee manoeuvre strategy. The main reason for this is that for the latter the apogee can be tailored to the particular transfer orbit obtained, and the error in the manoeuvres is essentially only the error of the final burn. These considerations require a completely different approach, and this is outlined below.

5.4.1 Statement of the Problem

Let us assume that the apogee manoeuvre consists of three burns. The long duration burns and their collective and individual optimisation make a full scale Monte Carlo analysis out of the question. For each simulation run, the optimisation would have to be performed, and this would render the analysis of the three burns and the subsequent station acquisition too costly.

The operational situation is as follows.

After launch, the spacecraft will be tracked to assess its orbit. This will be somewhat different from the nominal transfer orbit. Based on the tracked orbit, and here tracking errors will be incurred, the apogee manoeuvre strategy will be optimised to achieve the target orbit and on-station longitude. In the absence of all further errors, the transfer orbit tracking errors are all that prevent perfect realisation of the target orbit. In reality, however, burn errors will occur, and the intermediate orbit will differ somewhat from the desired one. A further period of tracking will determine the intermediate orbit to within the accuracy of the tracking procedure. Based on this information the optimisation will be performed again to size the two remaining burns to achieve the target orbit and on-station longitude. In the absence of all further errors, the intermediate orbit tracking errors are all that prevent perfect realisation of the target orbit. Now, however, some non-nominal fuel usage will have occurred through having had to re-optimise the second and third burns.

Burn errors in the second burn will result in a somewhat different intermediate orbit from the desired one. A further period of tracking will determine the intermediate orbit to within the accuracy of the tracking procedure. Based upon this data the final burn will be sized to achieve the target orbit and arrive as close as possible to the on-station longitude. Once more some non-nominal fuel use will have occurred through having had to re-target the final burn.

At the end of the final burn, burn errors will result in the final orbit being somewhat different from the desired one, and the longitude being other than the on-station one. Hence some station acquisition manoeuvres will be necessary. Taking into account attitude acquisition manoeuvres and the time limit on station acquisition, a station acquisition strategy will need to be determined and implemented. It can be seen that non-nominal fuel usage occurs for the apogee manoeuvre through the need to compensate for transfer orbit dispersions and tracking and burn errors, and non-nominal fuel usage occurs for the spacecraft's reaction control thrusters through having to perform station acquisition from a non-nominal starting condition.

As accurate modelling is not practicable, some sensible simplifications must be considered. Section 5.4.2 discusses a method based on reasonable simplifications.

5.4.2 The Method of Solution

Step 1

For a large number (~2000) of random and tracked transfer orbits, we generate apogee maneouvre ΔVs for the selected three-burn strategy assuming impulsive burns and ignoring orbital perturbations. (Note that this is equivalent to considering a single impulsive burn, because it is the total ΔV rather than the details of the individual burns that we are considering.) This gives the 99% ΔV and corresponding propellant mass for the apogee manoeuvre, yielding the non-nominal propellant requirements to counter transfer orbit dispersions and initial tracking errors. The calculation of the propellant mass from the ΔV must take into account specific impulse error, but not pointing error.

Step 2

Next we need to assess the influence of pointing errors during the apogee manoeuvres. Owing to the fact that the total burn is divided into three, pointing errors in the first burn can be countered by the second burn essentially without loss of ΔV, and the same can be said for the second burn. Burn errors in the final burn, however, could be expected to lead to a ΔV penalty for the subsequent station acquisition manoeuvres. The problem, therefore, reduces to that of assessing the third burn and subsequent station acquisition.

Now, the third burn is generally the smallest of the three and is made in an orbit with a near-synchronous semi-major axis. It is modelled as a long duration burn, with representative thrust vector steering.

We use nominal initial orbit conditions, which allow a definition of burn size and direction. We generate random pointing and steering errors from known pointing and steering performance capability and calculate the post-firing orbit. These orbits provide starting data for the station acquisition calculations in Step 3.

Step 3

Using the random drift orbits, and tracking errors in drift orbit, we evaluate the ΔV required for station acquisition. For this the algorithm described in Section 5.3, can be used. The time limit allowed for station acquisition must be respected, as must other operational constraints.

This analysis will yield the 99% ΔV required for station acquisition.

Synthesis

As observed earlier, the total non-nominal ΔV will arise in two components, that pertaining to the apogee manoeuvre calculated in *Step 1*, and the station acquisition ΔV calculated in *Step 3*. These two *could* be root-sum-squared to give the total non-nominal ΔV. However, because of the simplifications involved, a conservative approach may be to add the two components, as there will be some correlation between launch vehicle dispersions and station acquisition propellant requirements.

This technique has been used for the OLYMPUS and INMARSAT spacecraft.

5.5 Station Initialisation

Acquisition of the on-station longitude is followed by station initialisation manoeuvres whose objective is to achieve a set of orbit parameters consistent with the station-keeping strategy for the mission. The initialisation manoeuvres may be regarded as the first set of station-keeping manoeuvres, and are discussed in this chapter only because they occur immediately after station acquisition, and because the on-station phase of the mission cannot be said to have started until they have been satisfactorily performed.

The orbit parameters to be achieved by initialisation depend upon the station-keeping strategy. If station-keeping is to be performed within tight North-South and East-West deadbands – e.g. $\pm 0.1^{\circ}$ in both directions – the orbit pole must be placed such that the inclination is close to the maximum allowed, and the right ascension of the ascending node is approximately 270°E of the Vernal Equinox. In addition, the orbit apse line must be adjusted to place the direction of perigee just behind the Sun direction. This will ensure that the satellite does not transgress its deadband soon after the start of on-station operations.

If the station-keeping strategy is one of free-drift in the North-South direction, – as for MARECS – the initial placement of the orbit pole must be very accurate or corrective manoeuvres will subsequently be required.

An understanding of initialisation requires an understanding of the perturbations that act upon a satellite in geostationary orbit. These, and the strategies to counteract them, form the subject of Chapter 6.

5.6 Ground Station Coverage

Ground station coverage in drift orbit is required in order to command and monitor the following events:

- de-spin,
- Sun acquisition,
- Earth acquisition,
- station acquisition,
- tracking.

Owing to the position of the apogee manoeuvre, and the characteristics of the drift orbit, ground station coverage is extensive if the transfer orbit network, or an important part of it, is retained for this phase of the mission. As the position of the apogee manoeuvre is selected to be both near the on-station longitude, and covered by more than one ground station, at least two-station coverage will be obtained during the initial part of the *drift* orbit phase. As the drift orbit has only a small drift rate relative to the Earth, two-station coverage should, in fact, be obtained for the greater part of the drift orbit phase. Some possibility exists, however, that due to errors in the apogee motor firing, high drift rates could result. The ground station coverage for these cases must be checked with reference to the station acquisition strategies. The longitude variation for each of the station acquisition simulations, carried out during the Monte Carlo analysis described in Section 5.3, is examined, and the most adverse case within the probability limit is checked for adequate ground station coverage.

Since the events listed above are initiated upon ground command, continuous coverage of the drift orbit from at least one ground station is essential. For example, the ECS drift orbit is always covered by at least two stations, and, on occasion, three-station coverage is available, see Figure 5.2.

5.7 Orbit Determination Accuracy Requirements

If the station acquisition manoeuvres are to be performed accurately and efficiently, good knowledge of the orbit characteristics is necessary. In this section we shall outline some of the considerations leading to tracking accuracy requirements during the drift orbit phase of the mission.

Immediately after completion of the apogee manoeuvre(s), tracking is begun to characterise the orbit. About 24 hours are generally allocated for completion of this task, this time being dictated by the attitude acquisition sequence (embracing both prime and back-up opportunities). The first station acquisition manoeuvre may, therefore, be initiated 24 hours after the apogee manoeuvre. It is important to be able to initiate station acquisition as soon as possible in order to reduce a possible high drift rate away from the operational longitude. When the station acquisition

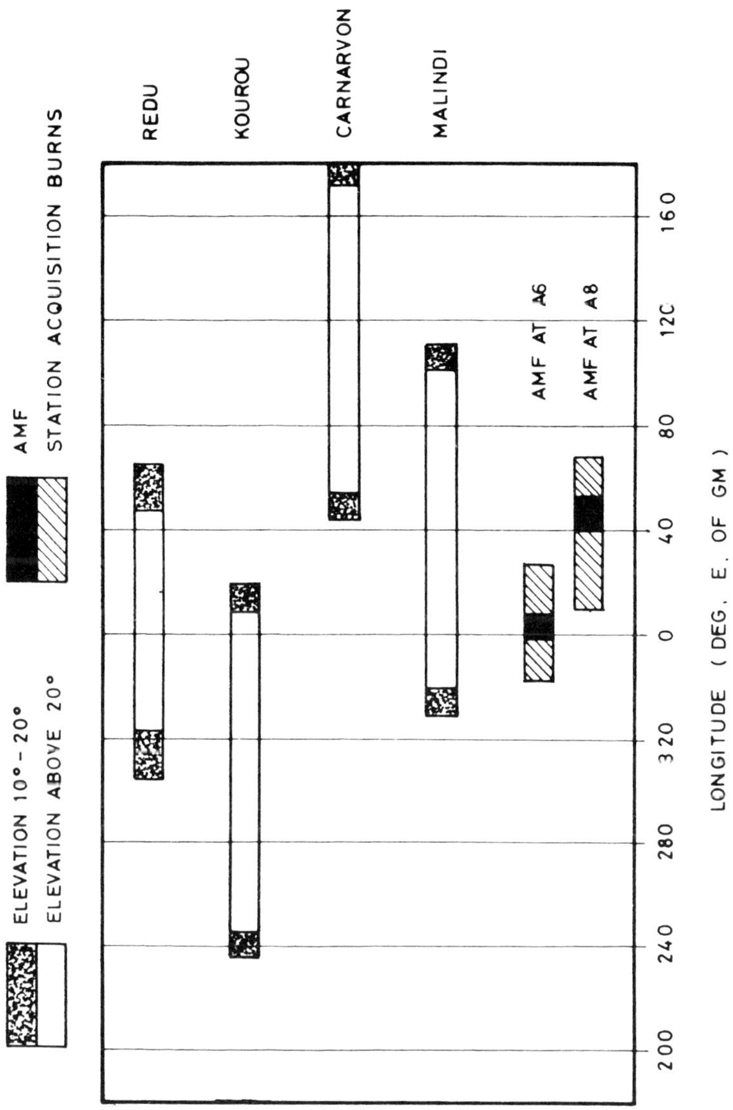

Fig.5.2 Ground Station Coverage for ECS in Drift Orbit

manoeuvre has been executed, tracking is undertaken again, with a view to enabling the subsequent manoeuvre within 24 hours. Upon completion of the final manoeuvre, the spacecraft will be within its station-keeping deadband. It must remain within this deadband until the orbit is determined accurately and station-keeping initialisation burns are performed. This period is generally two days. This determines the minimum drift rate accuracy required after the final station acquisition burn, and hence the orbit determination accuracy of the semi-major axis of the orbit.

Since the station-keeping initialisation burns make the fine adjustment to the orbit parameters, these manoeuvres make the greatest demands upon orbit determination accuracy. They lead to required accuracies on the following orbit parameters:

- pole position : characterised by inclination, i, and right ascension of the ascending node, Ω;
- apse heights and position : characterised by semi-major axis, a, eccentricity, e, and argument of perigee, ω;
- spacecraft longitude, λ, and drift rate : the latter characterised by the semi-major axis.

For ECS, the station-keeping dead-bands are ±0.1° in both the East-West and North-South directions. Considerations such as the above led to the following allowable maximum tracking error standard deviations:

$$
\begin{aligned}
\sigma(i) &\leqslant 0.005^{\circ} \\
(\sin i)\,\sigma(\Omega) &\leqslant 0.005^{\circ} \\
\sigma(r) &\leqslant 1.67 \text{ km} \\
e\,\sigma(\omega) &\leqslant 0.002^{\circ} \\
\sigma(\lambda) &\leqslant 0.003^{\circ} \\
\sigma(a) &\leqslant 0.1 \text{ km}
\end{aligned}
$$

where $\sigma(x)$ = standard deviation of x, and r = apse radius

For MARECS with a North-South free drift strategy, an initial pole position of $i = 3^{\circ}$ and $\Omega = 299^{\circ}$, and an East-West deadband of ±0.5°, the following requirements arose:

$$
\begin{aligned}
\sigma(i) &\leqslant 0.01^{\circ} \\
(\sin i)\,\sigma(\Omega) &\leqslant 0.01^{\circ} \\
e\,\sigma(\omega) &\leqslant 0.006^{\circ} \\
\sigma(\lambda) &\leqslant 0.1^{\circ} \\
\sigma(a) &\leqslant 0.8 \text{ km}
\end{aligned}
$$

Note that the latter requirements are less stringent owing to the larger station-keeping deadband, and the available tolerance on the location of the initial pole position.

It should be noted that manoeuvre and tracking accuracy requirements are closely dependent upon the capabilities of the ground control station and associated network. For example, the minimum time between manoeuvres is determined by ground station procedures. As such operational constraints have an influence upon the station acquisition propellant requirement and tracking schedules, they must be known during the mission design phase, i.e. many years before the mission is actually flown.

5.8 Further Reading

The Monte Carlo station–acquisition analysis is discussed in the paper below:

(a) Rao, P.P. : 1978, "Monte Carlo Analysis of Impulse Requirements for Injection Error Correction", *J. of Guidance and Control*, Vol.1, No.4, pp.225–231.

General mission design for spacecraft with liquid propellant engines is discussed in:

(b) Pocha, J.J. : 1984, "Mission and System Analyses for Satellites with Unified Bi-Propellant Propulsion Systems", *Space Communications and Broadcasting*, Vol.2, pp.59–70.

A discussion of an analytic approach to the error analysis using the covariance matrix method is given in:

(c) Porcelli, G. and Vogel, E. : 1980, "Two-Impulse Transfer Error Analysis via Covariance Matrix", *J. of Spacecraft*, Vol.17, May–June, pp.248–255.

CHAPTER 6

STATION-KEEPING

6.1 Introduction

A spacecraft in geostationary orbit changes its position with respect to an Earth-fixed frame of reference owing to the action of a number of perturbations. The word *perturbation* is used to signify forces other than those due to the gravitational potential of a homogeneous, spherical Earth. The effect of the perturbations is to cause the spacecraft to drift away from its nominal station. If this drift was allowed to build up unchecked, the spacecraft could no longer be used as a communications relay station. *Station-keeping*, therefore, has to be instituted, and the spacecraft is manoeuvred in order to keep it within strict latitude and longitude limits defining a *dead-zone*.

The magnitude of the dead-zone depends primarily upon the characteristics of the communications transponder. For example, satellites employing antennae providing global coverage may be maintained within a fairly large dead-zone, e.g. $\pm 0.5^{\circ}$ of latitude and longitude, while those employing narrow-beam high-gain antennae must be kept within much stricter limits, e.g. $\pm 0.1^{\circ}$. The latter is typical of the ECS family of communications satellites, and is representative of most regional communications spacecraft. With increasing demand for positions on the geostationary arc, spacecraft are being maintained within ever tighter dead-zones.

In order to devise efficient and operationally convenient station-keeping strategies, we must study the nature of the perturbations, and their effect upon spacecraft kinematics. Fortunately, to sufficient accuracy, the perturbations dealing with longitude and latitude drift arise from separate physical mechanisms, and can, therefore, be de-coupled and studied separately. It is only when devising manoeuvre strategies to counteract these perturbations that some cross-coupling must be taken into account. For this reason we shall consider the East-West and North-South cases separately, and highlight their interactions.

6.2 East-West Station-Keeping

A geostationary satellite drifts in longitude due to the influence of two main perturbations. The first one arises from the non-sphericity of the Earth, and hence is caused by higher order gravitational harmonics. The second arises from the effect of solar radiation pressure on the satellite's sun-facing surfaces (these would be mainly the large solar arrays typical of communications satellites). As the two perturbations are independent of each other they are discussed separately.

6.2.1 Triaxiality

6.2.1.1 The Physical Mechanism

The perturbation due to the non-sphericity of the Earth arises from the elliptic nature of the Earth's equatorial cross-section. The Earth's oblateness, in addition to the elliptic equatorial cross-section results in our being able to describe the Earth, to fair accuracy, as an ellipsoid with three dissimilar axes. Actually, the perturbations are due to the complete set of harmonics in the Earth's gravitational field, but at synchronous altitude the effects of harmonics other than the one associated with triaxiality (the harmonic of order $m = n = 2$, or J_{22}) are small.

The effect of triaxiality can be considered to be a small transverse acceleration on the spacecraft, as shown in Figure 6.1. The magnitude of the transverse acceleration depends upon the location of the satellite with respect to the major and minor axes of the equatorial ellipse, i.e. upon satellite longitude. Owing to the fact that at synchronous altitude the satellite's orbit period is closely matched to the Earth's rotation rate, the satellite geographic longitude is constant or changes only slowly. The triaxiality perturbation, therefore, results in a near constant transverse acceleration upon the satellite. The near coincidence of the orbit period and the Earth's rotation period has lead to this perturbation also being known as the *resonance effect*.

The magnitude of the acceleration is given by:

$$f_\lambda = - \frac{1}{r \cos L} \quad \frac{\delta u}{\delta \lambda}$$

where:
λ = satellite longitude,
L = satellite latitude,
r = radial distance,
u = perturbing potential due to Earth harmonics.

The transverse acceleration can be expressed as a velocity change per year, and in this form is shown in Figure 6.2, where we have two curves, one showing the perturbation due to triaxiality alone, the other showing that due to all the harmonics (in practice up to about $m = n = 4$). The longitudes marked S_1 and S_2 indicate stable longitude positions, while those marked U_1 and U_2 indicate unstable positions. To understand the significance of these positions we consider the long period motion of the satellite under the influence of triaxiality perturbations.

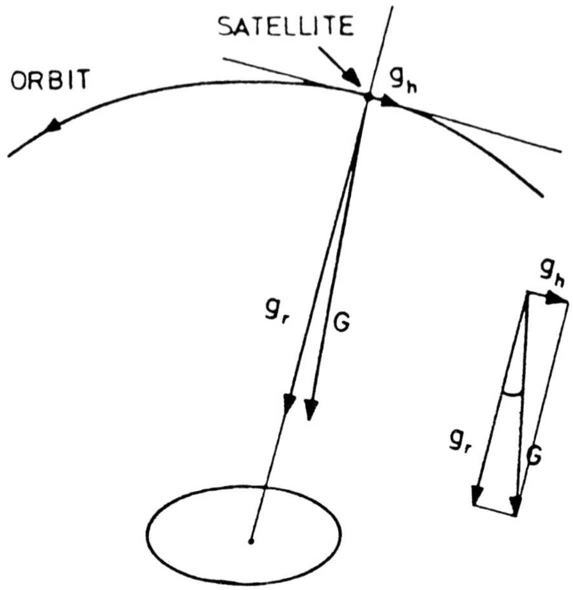

Fig.6.1 The Triaxiality Acceleration on the Satellite

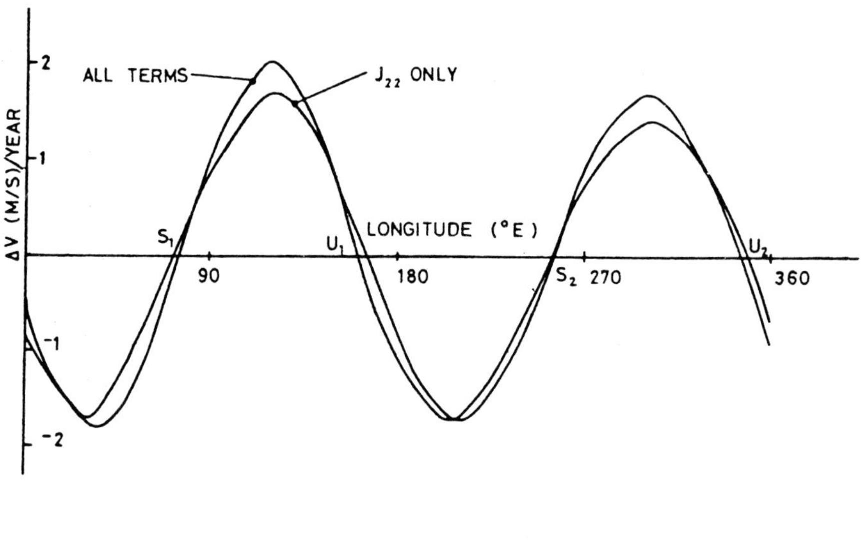

Fig.6.2 The Velocity Increment for Triaxiality Compensation

If the satellite is at some longitude λ_1 and the perturbations are not counteracted by manoeuvres, the satellite will drift under the influence of the transverse acceleration. See Figure 6.3. The transverse acceleration will retard the satellite and cause it to spiral downwards. As it does so the orbit period will decrease, and the satellite will seem, to an Earth-fixed observer, to drift eastwards. When the satellite reaches a longitude λ_2 the transverse acceleration will become zero. Any further movement of the satellite will cause the transverse acceleration to tend to speed up the satellite and the situation will begin to reverse itself. The satellite will gradually gain altitude and the eastward drift will slow down until at point λ_3 it is zero. The satellite will now begin to drift westwards back to the initial position. The period of this motion is of the order of 800 days. In Figure 6.3 longitude λ_2 is the stable point at which the perturbation is zero, and at which the long period motion is centred. The point is coincident with the semi-minor axis of the equatorial ellipse. The unstable points are coincident with the semi-major axis. Long period motion can never be centred about an unstable point as *apparent* acceleration is always away from it.

Clearly, if the satellite was allowed to drift freely under the effect of triaxiality, it would soon drift far away from its desired position, and become useless as a communications platform. Its longitude must therefore be controlled by periodic manoeuvres, and be maintained close to some desired value. Because this is the case, we can consider the transverse acceleration due to triaxiality perturbations to be constant. To maintain station we must counteract this transverse acceleration by means of on-board thrusters, and the required annual velocity increment, ΔV, is given by Figure 6.2. It depends only on satellite longitude.

6.2.1.2 Satellite Kinematics

In order to devise a strategy to maintain the satellite's longitude within the desired limits, we must now consider satellite kinematics. For this we make use of the *Variation of Parameters* equations and assume that the nominal orbit is a geostationary one, and that the perturbing transverse acceleration, g_h, is very much smaller than the radial gravitational acceleration g_r. Refer to Figure 6.1. We have that:

$$g_h << G; g_h << g_r; \theta << 1$$

According to the *Variation of Parameters* equations, the change in semi-major axis, Δa, due to a transverse acceleration, g_h, is given by:

$$\frac{da}{dt} = \frac{2pg_h}{nr\,(1-e^2)^{1/2}}$$

where:
n = orbit rate, also called the *mean motion*,
p = $a\,(1-e^2)$,
e = eccentricity,
r = radius,
a = semi-major axis of the orbit.

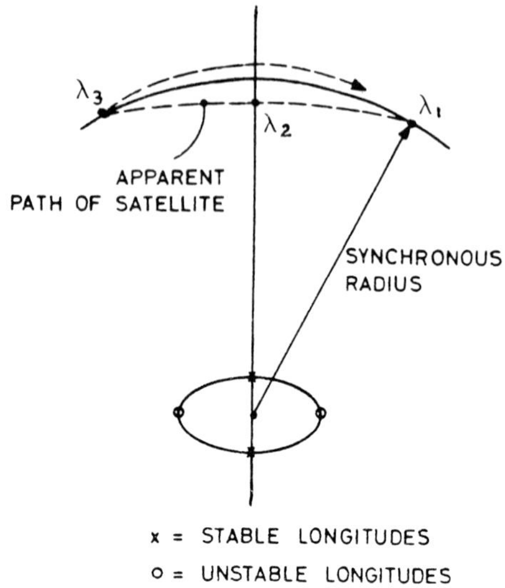

Fig.6.3 The Satellite's Apparent Motion Due to Triaxiality

For geostationary orbits this becomes:

$$\frac{da}{dt} = \frac{2\,g_h}{n}$$

$$\therefore \qquad \Delta a = \frac{2\,g_h\,\Delta t}{n}$$

Now the change in semi-major axis, in the time it takes for the satellite to move round the orbit by 1 radian, is:

$$\Delta a = \frac{2\,g_h}{n^2} = \frac{2\,g_h\,a^3}{\mu}$$

where: $n = \sqrt{\mu/a^3}$,
μ = Earth's gravitational constant.

The acceleration due to a spherical Earth is given by:

$$g_r = \frac{\mu}{a^2}$$

$$\therefore \qquad \Delta a = \frac{2ag_h}{g_r} = 2a\theta$$

where: $\theta = g_h/g_r$

The new semi–major axis, a', is given by:

$a' = a(1 + 2\theta)$ for g_h in the direction of motion.

The new velocity in the orbit is given by:

$$V' = \left(\frac{\mu}{a'}\right)^{\frac{1}{2}} = \frac{\mu^{\frac{1}{2}}}{a^{\frac{1}{2}}(1 + 2\theta)^{\frac{1}{2}}} = \frac{V}{(1 + 2\theta)^{\frac{1}{2}}}$$

Neglecting terms of order θ^2 :

$$V' = V(1-\theta)$$

The orbital angular rate, ω', is given by:

$$\omega' = \frac{V'}{r'} = \frac{V(1-\theta)}{a(1 + 2\theta)}$$

$$\therefore \qquad \omega' = \omega(1-\theta)(1-2\theta) = \omega(1-3\theta)$$

Hence the angular motion of the satellite *relative to the Earth* is $-3\theta\omega$ rad/sec.

Now $$\frac{d\omega}{df} = -3\theta\omega \text{ (rad/sec)/radian}$$

$$\frac{d\omega}{dt} = \frac{d\omega}{df} \cdot \frac{df}{dt} = -3\theta\omega \,.\, n = -\frac{3\theta}{a} \cdot \sqrt{\frac{\mu}{a}} \cdot \sqrt{\frac{\mu}{a^3}}$$

$$\therefore \qquad \dot{\omega} = -\frac{3\theta}{a}\frac{\mu}{a^2}$$

But $(\mu/a^2) = g_r$ for small θ

$$\therefore \qquad \dot{\omega} = \frac{3\theta g_r}{a}$$

Hence the apparent linear acceleration is given by:

$$a\dot{\omega} = -3\theta g_r = -3g_h$$

This is an important result, and indicates that the *apparent* acceleration is –3 times the perturbing transverse acceleration, and because the latter is taken to be constant, the *apparent* acceleration is also constant.

If we work in terms of angular position, or longitude, λ, and angular rates, the *apparent* angular acceleration is $\dot{\omega} = -3g_h/a$ = constant. Let us denote this as $\ddot{\lambda}$ = constant.

Hence we have that:

$$\dot{\lambda} = \dot{\lambda}_0 - \ddot{\lambda} t$$

and

$$\lambda = -\lambda_0 + \dot{\lambda}_0 t - \frac{\ddot{\lambda}}{2} t^2$$

These are our kinematic relationships.

We can now devise a manoeuvre strategy that will keep the satellite within the stipulated deadband, and also minimise the number of manoeuvres to do so. The latter is an important consideration for satellite operations, as the execution of manoeuvres requires ground station resources, and the satellite reliability decreases for an increase in the number of thruster firings.

The following strategy, therefore, suggests itself. We set up the satellite at one end of the deadband and give it an initial drift rate such that the apparent acceleration just reduces it to zero by the time the satellite reaches the opposite edge of the deadband. At this point the satellite will begin to move towards the centre of the deadband. At some time it will overshoot the nominal longitude (at the centre of the deadband) and move towards the starting position. When this is reached, a manoeuvre reverses the drift rate and the cycle begins again.

With such a method, we make the maximum use of the deadband and obtain the maximum time between corrections. This is the same as minimising the number of manoeuvres for a certain life-time. The strategy described above is illustrated on the λ, $\dot{\lambda}$ plot of Figure 6.4.

The satellite starts at point A at one edge of the deadband, with an initial drift rate of $\dot{\lambda}_0$. By the time it reaches C, the opposite edge of the deadband, its drift rate has been reduced to zero. The satellite then drifts towards B. When it reaches B a manoeuvre is executed to change its drift rate from $-\dot{\lambda}_0$ to $+\dot{\lambda}_0$. After the manoeuvre, the satellite is again at A.

It is interesting to note that it is not the position, to which the deadband applies, but the drift rate that is the subject of the correction manoeuvre. This principle will appear again when we discuss solar radiation pressure perturbations.

The cycle time for the above strategy can be derived as follows. See Figure 6.4.

At C:

$$\lambda_0 = -\lambda_0 + \dot{\lambda}_0 t_0 - \frac{\ddot{\lambda}}{2} t_0^2$$

and

$$0 = \dot{\lambda}_0 - \ddot{\lambda} t_0$$

$\therefore$

$$\dot{\lambda}_0 = \ddot{\lambda} t_0$$

and

$$2\lambda_0 = \ddot{\lambda} t_0^2 - \frac{\ddot{\lambda}}{2} t_0^2 = \frac{\ddot{\lambda}}{2} t_0^2$$

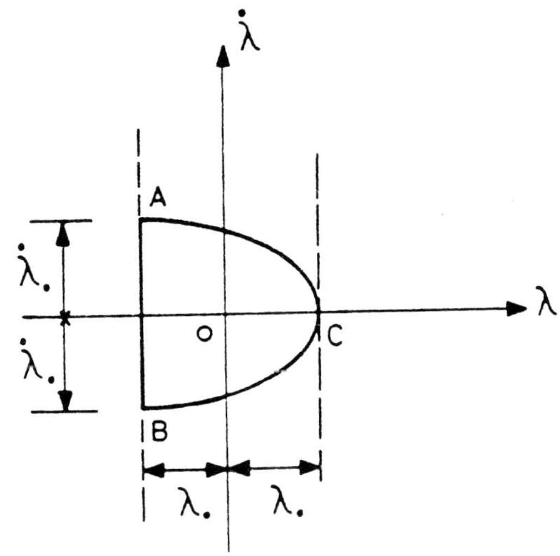

Fig.6.4 The Triaxiality Perturbation Correction Strategy

which gives:

$$\therefore \qquad t_o = 2\sqrt{\frac{\lambda_o}{\ddot{\lambda}}}$$

$$\text{and} \qquad \dot{\lambda}_o = 2\sqrt{\ddot{\lambda}\,\lambda_o}$$

$$\therefore \qquad \text{Cycle time} = 2\,t_o = 4\sqrt{\frac{\lambda_o}{\ddot{\lambda}}}$$

Since the manoeuvre consists of a change in the drift rate by $2\dot{\lambda}_o$, this is equivalent to a change in the semi-major axis. To accomplish this without changing the eccentricity requires a minimum of two burns, 180° apart. Both burns are made in the same direction relative to the velocity vector. The first burn changes the semi-major axis by half the required amount, and sets up an eccentricity. The second burn completes the semi-major axis change and cancels the eccentricity set up previously.

We can calculate the change in semi-major axis required to change the drift rate from the equation:

$$\Delta\dot{\lambda} = -\frac{3n\Delta a}{2\,a_G}$$

where: n = geosynchronous inertial orbit rate,
= 360.986°/day,
a_G = geosynchronous radius.

$$\therefore \quad \Delta a = \frac{2}{3}\,(2\dot{\lambda}_0)\,\frac{a_G}{n}$$

The burn magnitude (ΔV) required to achieve this change in the semi-major axis is given by:

$$\Delta V = \frac{V_G \Delta a}{2a_G}$$

where: V_G = geosynchronous orbit velocity,
a_G = geosynchronous orbit radius.

Alternatively the burn magnitude can be obtained from Figure 6.2 by dividing the ΔV/year, by the number of manoeuvres required through 1 year. It can be shown that the result will be exactly the same.

It should be noted that for triaxiality corrections, the total expenditure of ΔV is not dependent upon the manoeuvre strategy. It was shown before that a correction manoeuvre required at least two impulses 180° apart. There may, if needed, be even more than two burns. As long as the impulses are tangential and have the same sense with respect to the velocity vector, and if their vectors form a closed figure, their aggregate effect is exactly the same as that of two impulses. The semi-major axis is changed, but the eccentricity remains unaffected. This timing insensitivity of triaxiality corrections is a most useful asset when combining the manoeuvre schedules of the various perturbation corrections.

6.2.2 Solar Radiation Pressure

6.2.2.1 The Physical Mechanism

A satellite exposed to the Sun suffers a radiation pressure which results in an acceleration upon the satellite in the Sun-satellite direction. The principal effect of solar radiation pressure on a geostationary satellite is to produce a long-period cyclic perturbation in the orbit eccentricity, while leaving the semi-major axis essentially unchanged.

For an orbit which is initially circular the mechanism by which an eccentricity is acquired is shown in Figure 6.5. The effect of the acceleration due to solar radiation pressure integrated over the lower half of the orbit may be represented by a small velocity increment (ΔV) adding to the circular velocity at point 1 and tending to raise the altitude (produce an apogee) at the diametrically opposite point. Similarly a small ΔV opposing the circular velocity at point 2 will represent the integrated effect over the upper half of the orbit. This will tend to lower the altitude (produce a perigee) at the diametrically opposite point. Thus, without changing the semi-major axis, an eccentricity is developed, and the associated line of apsides is

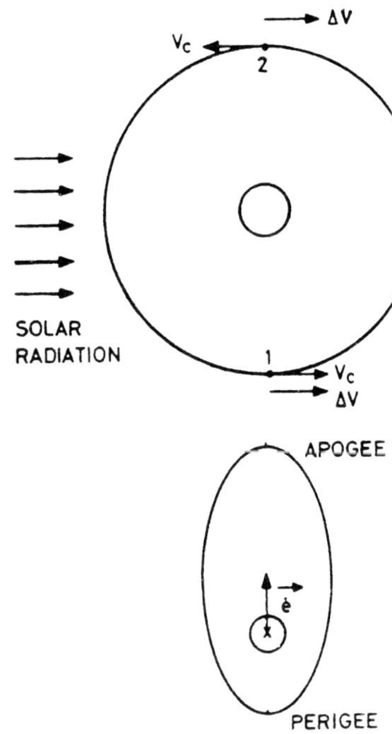

Fig.6.5 The Solar Pressure Perturbation

normal to the Sun line. As the Sun direction changes, the line of apsides rotates in an Earth–fixed frame of reference.

At this point we introduce the concept of the *eccentricity vector* defined as a vector in the direction of the apogee, of a magnitude equal to the eccentricity of the orbit. We now make some simplifying assumptions in order to make the analysis tractable. We assume that:

- the satellite area–to–mass ratio and reflectance are constant;
- the satellite is never eclipsed;
- the Sun–Earth distance is constant;
- the equator is co–planar with the ecliptic;
- there are no interactions with other orbit perturbations.

With these assumptions, we find that the locus of the centre of the elliptic orbit is a circle with the motion of the centre perpendicular to the Sun line at all times. If the orbit is initially circular, the orbit centre begins at the occupied focus as shown in Figure 6.6.

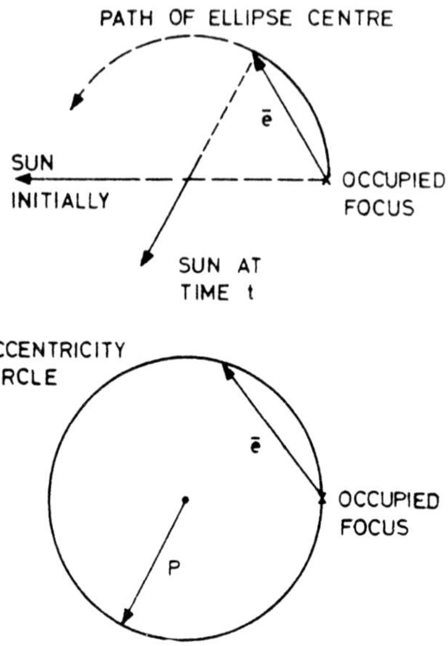

Fig.6.6 Evolution of the Eccentricity Vector

Since the actual distance between the occupied focus and the ellipse centre is equal to e (for the semi-major axis as a unit), this plot may be used to determine the variation of the eccentricity vector. The locus of the tip of the eccentricity vector is known as the *eccentricity circle*. A full circle is described over a period of one year.

6.2.2.2 Satellite Kinematics

The effect of solar radiation pressure, to first order, is to generate an eccentricity while leaving the semi-major axis unchanged. We shall use the *Variation of Parameters* equations to derive the rate of change of eccentricity. Refer to Figure 6.7.

The satellite experiences a constant acceleration, F, acting along the Sun–satellite direction.

$$\begin{aligned} F &= T\cos f + S\sin f \\ 0 &= -T\sin f + S\cos f \end{aligned} \qquad (1)$$

where: f = true anomaly,
S = radial acceleration,
T = transverse acceleration.

We shall assume that eccentricity, e, is always small, i.e. $e \ll 1$.

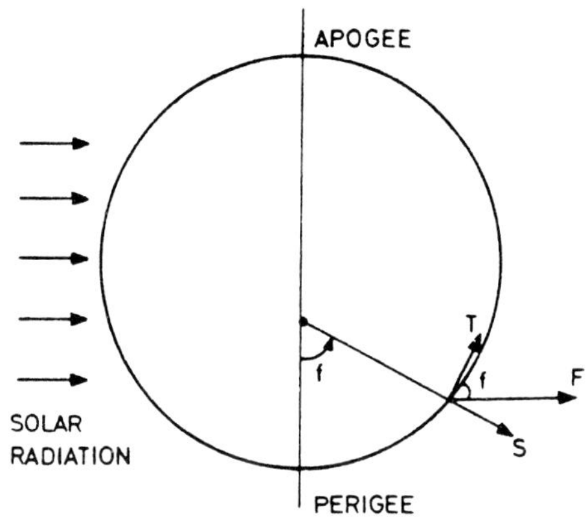

Fig.6.7 Force on the Satellite Due to Solar Pressure

From the Variation of Parameters equations, we have:

$$\frac{de}{dt} = \frac{(1-e^2)^{1/2}}{na} \quad [S \sin f + T (\cos E + \cos f)]$$

where: e = eccentricity,
a = semi–major axis,
n = mean motion,
f = true anomaly,
E = eccentric anomaly.

For $e \ll 1.0$, this reduces to:

$$\frac{de}{dt} = \frac{1}{na} \quad (S \sin f + 2\, T \cos f)$$

Using Eqn. (1) we get:

$$\frac{de}{dt} = \frac{1}{na} \quad (2F - S \sin f) \quad = \frac{1}{V_c} \quad (2F - S \sin f)$$

where $na = V_c$ = velocity in circular orbit.

$$S = F \sin f$$

$$\therefore \qquad \frac{de}{dt} = \frac{1}{V_c}\,(2F - F\sin^2 f) \qquad (2)$$

Now let the change in eccentricity per orbit (or per day) be Δe.

$$\Delta e = \int_{t=0}^{t=T} \frac{de}{dt}\,dt$$

Using Equation (2),

$$\Delta e = \int \left(\frac{2F}{V_c} - \frac{F}{V_c}\sin^2 f\right) dt = \frac{2F}{V_c} - \frac{F}{V_c}\int \sin^2 f\,dt$$

Now $\displaystyle\int_0^T \sin^2 f\,dt = \int_0^{2\pi} \frac{\sin^2 f\,df}{df/dt} = \frac{1}{n}\int_0^{2\pi} \sin^2 f\,df$

$n = 2\pi$ rad/day for a geosynchronous orbit.

$$\therefore \qquad \frac{1}{n}\int_0^{2\pi} \sin^2 f\,df = \frac{1}{2\pi}\cdot \pi$$

$$\therefore \qquad \Delta e = \frac{2F}{V_c} - \frac{F}{2V_c} = \frac{3}{2}\cdot\frac{F}{V_c} \quad \text{(per day)}$$

The direction of Δe is always normal to the Sun direction.

Now, the acceleration, F, due to solar radiation pressure is given by the expression:

$$F = G\left(\frac{A}{M}\right)\left(\frac{1+R}{2}\right) = G\left(\frac{A}{M}\right)_{\text{eff}}$$

where: G = solar radiation pressure for a perfectly reflecting flat surface,
= 9.1×10^{-6} N/m^2,
A = satellite area projected normal to the Sun line, (m^2),
M = satellite mass (kg),
R = surface reflectance,
= 1 for perfect reflector,
= 0 for perfect absorber.

From the above, it can be seen that for most satellites, to a very good approximation, the rate of change of eccentricity is constant.

Owing to the constancy of Δe, the locus of the tip of the eccentricity vector will, in the absence of corrective manoeuvres, describe a circle in the course of a year. This circle is the *eccentricity circle*.

The radius of the eccentricity circle, p, is given by:

$$p = \frac{365.25}{2\pi} \cdot \dot{e} = 0.0223 \left(\frac{A}{M}\right)_{\text{eff}}$$

where (A/M) eff is expressed in (m^2/kg).

Since the eccentricity circle retains its size and shape while being translated for starting conditions other than circular orbits, we can see that if $e_0 = p$ and the perigee is towards the Sun, the eccentricity will remain constant and equal to p. Furthermore, the eccentricity will be only half of the maximum eccentricity occurring for the initially circular case.

When a geosynchronous orbit has a non-zero eccentricity, e, the satellite longitude moves with a periodic (daily) motion relative to an Earth-bound observer. This motion is called *libration*, and the amplitude of the libration can be shown to be $2e$ radians. This can be shown simply as follows. See Figure 6.8.

When the satellite is at apogee or perigee it has a maximum drift rate relative to an Earth-bound observer, while at the edges of the libration band its drift rate is zero.

$\therefore$ At B and C, $r = a$, the semi-major axis.

Now consider the elliptic orbit formula for r, the radius vector:

$$r = \frac{a\,(1 - e^2)}{1 + e \cos f}$$

To obtain the true anomaly at B or C we have:

$$a + a\,e \cos f = a - ae^2$$

$$\therefore \quad \cos f = -e$$

From Figure 6.8 it can be seen that at C

$$f = \pi/2 + e\,.$$

We now obtain the value of the corresponding mean anomaly, M, as it is representative of the angle in the corresponding circular orbit.

$$M = E - e \sin E$$

where E = eccentric anomaly.

Now: $\quad r \cos f = a\,(\cos E - e)$

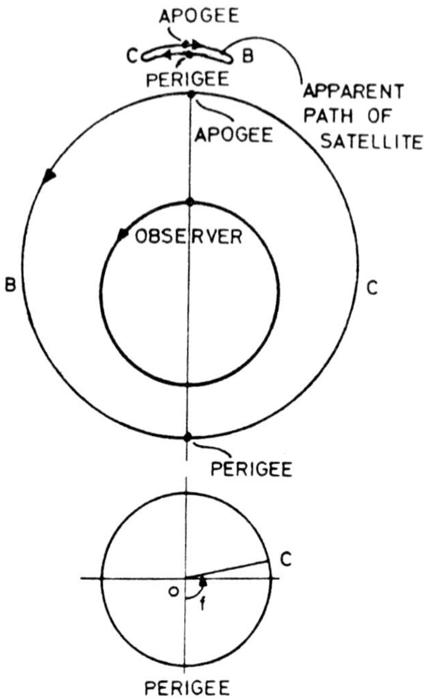

Fig.6.8 The Satellite's Apparent Motion Due to Solar Pressure

But at C, $\cos f = -e$ and $r = a$

$$\therefore \quad -ae = a \cos E - ae$$

$$\therefore \quad \cos E = 0 \text{ and } E = \frac{\pi}{2}$$

$$\therefore \quad M = \frac{\pi}{2} - e \text{ radians}$$

The difference between f and M is the change in the longitude of the satellite, $\Delta\lambda$, as seen by an Earth-fixed observer.

$$\Delta\lambda = f - M = \frac{\pi}{2} + e - (\frac{\pi}{2} - e)$$

$$\therefore \quad \Delta\lambda = 2e$$

Part of the satellite's longitude deadband must be allocated to libration motion. The extent of the allocation will determine the maximum eccentricity the orbit can be permitted to have. Hence the limitation on longitude reduces to a limitation on eccentricity. It can be seen that if the eccentricity tolerance, e_{tol}, is greater than or equal to $2p$ no manoeuvres are needed for an initially circular orbit. If $2p > e_{tol} > p$, manoeuvres can still be avoided by selecting the initial conditions so as to centre the eccentricity circle at the occupied focus. If, however, $e_{tol} < p$, correction manoeuvres become necessary.

In devising a correction strategy, two considerations assume importance. These are, the propellant required, and the number of manoeuvres needed. If the number of manoeuvres is minimised, the propellant required will be maximised, and vice versa. In practice the manoeuvre execution time could be dictated by requirements other than those solely related to solar radiation pressure corrections. However, we shall consider first the strategy to minimise the number of manoeuvres.

With such an objective, the strategy is to set the initial eccentricity vector such that the maximum time elapses before a correction is required. If the initial conditions are arranged such that the eccentricity is set at the maximum permitted value and the perigee is placed behind the Sun line, the perturbations will cause the eccentricity vector to begin decreasing in magnitude initially. It will reach some non-zero minimum, and then increase again. When the eccentricity has reached the maximum allowed value, a correction manoeuvre will be performed to shift the line of apsides to the starting position with respect to the Sun line, while leaving the eccentricity unchanged. Hence, it is not the eccentricity, upon which the tolerance is placed, that is subject to correction, but the direction of the line of apsides. This is, in principle, similar to the triaxiality correction where it is not the longitude but the drift rate that is corrected.

The strategy is illustrated in Figure 6.9. To maximise the time between corrections we must maximise the dwell time within the tolerance limit. Hence we maximise the arc length within the tolerance circle. The tip of the eccentricity vector moves from one point on the tolerance circle, to a point diametrically opposite.

At the start of the cycle the line of apsides is at an angle $\Delta\omega_0$ behind the Sun line.

$$\Delta\omega_0 \quad = \quad \cos^{-1}\,(e_{tol}/p)$$

At the end of the cycle the line of apsides leads the Sun line by $\Delta\omega_0$. The correction manoeuvre to re-establish the initial conditions must, therefore, move the line of apsides back through an angle of $2\Delta\omega_0$ radians. The time between corrections is the time taken for the Sun line to move through an angle of $2\,\sin^{-1}\,(e_{tol}/p)$.

Hence: $$\text{Cycle time} \;=\; t_c \;=\; \frac{1}{\pi}\,\sin^{-1}\,(e_{tol}/p)\;(\text{years})$$

The correction strategy has now been established. In order to move the line of apsides through $2\Delta\omega_0$ the optimum manoeuvre consists of two equal tangential burns located approximately along the line joining the vacant foci of the pre- and post-correction orbits. The first impulse is retrograde, and the second is prograde, as shown in Figure 6.10. The order in which the burns are performed is immaterial. In order to keep the satellite libration centred on the nominal longitude the burns must be executed symmetrically, e.g. two retrograde impulses of ½ magnitude 24 hours apart with one prograde impulse of full magnitude in between.

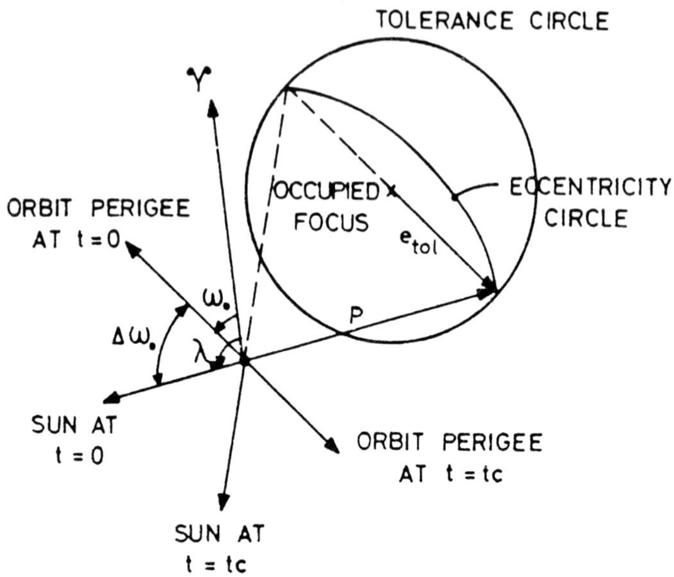

Fig.6.9 The Correction Strategy for Maximum Cycle Time

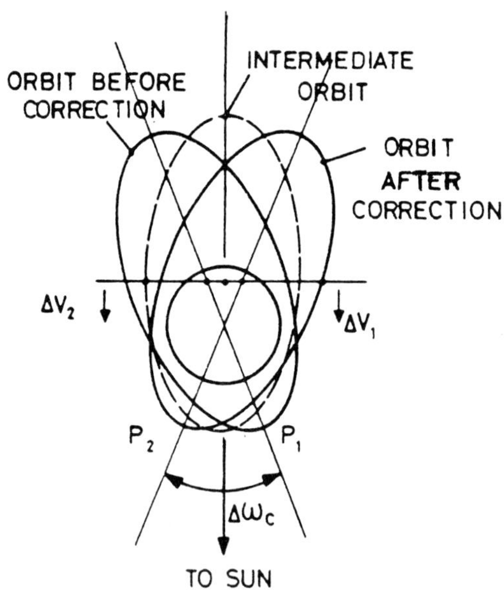

Fig.6.10 Solar Pressure Correction Burns

The total correction ΔV is given by:

$$\Delta V = V_c e_{tol} \sin \Delta\omega.$$

The total annual ΔV is given by:

$$\Delta V \text{ (annual)} = \Delta V / t_c$$

We now consider the strategy to minimise the total correction velocity requirement, i.e. we minimise ΔV. If the orbit initial conditions cannot be arranged to eliminate manoeuvres completely, the ΔV can be minimised by keeping the perigee directed very nearly towards the Sun, by making *frequent* corrections. In the limit, daily corrections are required but the penalty for rather less frequent corrections is small.

In general, the strategy is as before, except that now the eccentricity vector does not move from one point on the tolerance circle to a point diametrically opposite. The situation is shown in Figure 6.11. At the end of the cycle, the perigee is as far ahead of the Sun line as it was behind at the start of the cycle. The correction manoeuvre once again restores the initial condition.

Let the desired cycle time = t_c (years)

From Figure 6.11:

$$\Delta\omega_0 = \sin^{-1}\left[\frac{p}{e_{tol}} \sin(\pi t_c)\right] - \pi t_c$$

The correction manoeuvre rotates the line of apsides through an angle $2\Delta\omega_0$. The ΔV required for the manoeuvres is given, as before, by:

$$\Delta V = V_c e_{tol} \sin \Delta\omega_0 .$$

The manoeuvres are carried out exactly as before. For synchronous satellites the eccentricity tolerance is usually very small i.e. $e_{tol} \ll 1$. In this case, to a very good approximation, the manoeuvres are carried out at 6 am and 6 pm solar time.

6.2.3 Manoeuvre Strategies

6.2.3.1 Synthesis of Manoeuvre Strategies

We have seen that both triaxiality and solar radiation pressure corrections require thrusting in the transverse direction. Furthermore, triaxiality corrections can be made anywhere in the orbit while solar pressure corrections must be made at specific positions in the orbit. This suggests that the two manoeuvres can usefully be combined.

If the ΔV required for a triaxiality manoeuvre is $2T$ and that for a solar radiation pressure manoeuvre is $2P$, the manoeuvres can be combined as shown in Figure 6.12. On one side of the orbit the combined manoeuvre is $(T + P)$, while on the other it is $(T - P)$. If $T > P$ the total manoeuvre ΔV is $2T$, and if $P > T$, the total

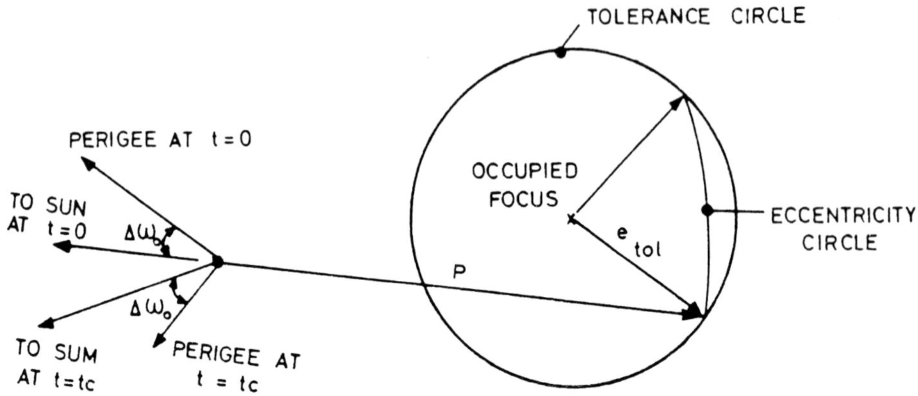

Fig.6.11 The Solar Pressure Correction Strategy

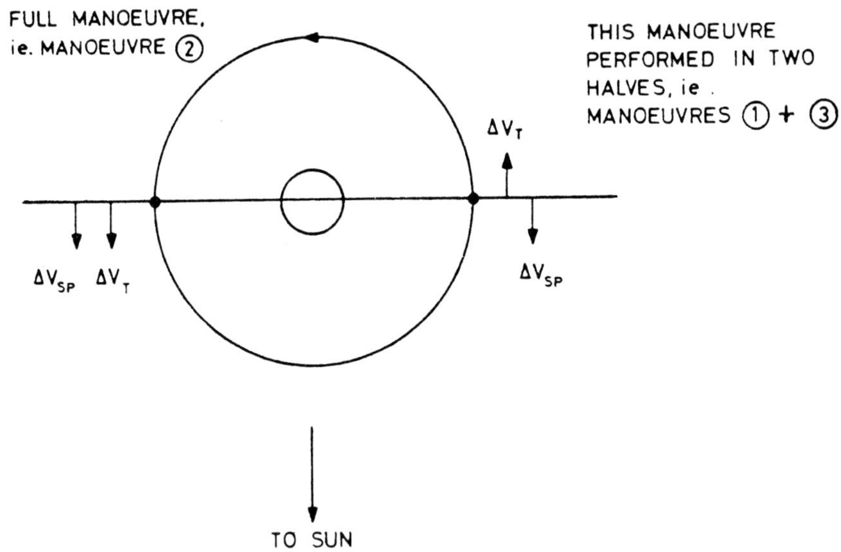

Fig.6.12 Synthesis of Triaxiality and Solar Pressure Corrections

manoeuvre ΔV is $2P$. For this to apply the manoeuvres must be executed concurrently, as shown in Figure 6.12. This means that the cycle times for both corrections must be the same. This must be taken into account when allocating the available longitude deadband between the two sources of perturbation.

The above strategy assumes that we correct for both the triaxiality and solar pressure effects to the full extent required by the cycle time. Other, simpler, strategies are also available and one such strategy is discussed next.

6.2.3.2 The One–Burn Strategy

This strategy, originated by Telesat, Canada, has been adopted for the present generation of European communications satellites, i.e. ECS and MARECS. A single burn strategy has operational attractions such as:

- the number of thruster firings is reduced and each firing is of a reasonably large magnitude to enable high total impulse accuracy to be obtained;
- the spacecraft is in station–keeping mode for a minimum of time thus minimising gyro switch–ons and warm–ups and the number of station–keeping–to–normal mode transient suppression manoeuvres;
- adequate tracking time is available before and after the burn;
- ground station work–load is minimised.

Conceptually, the method is more complex. The burn magnitude and application point are chosen to achieve the desired drift rate correction and at the same time to obtain a certain degree of eccentricity vector change. The method uses only the fuel required for triaxiality corrections. Triaxiality effects are compensated for completely, i.e. the burn is sized to achieve the correct drift rate. The timing of the manoeuvre is used to produce a parasitic eccentricity vector change. The basis of the method is that the orbit perigee is made to track the Sun as closely as possible, and the initial eccentricity is adjusted to suit the strategy. This keeps the eccentricity within reasonable limits for only the expenditure of the fuel necessary to correct triaxiality. The method is only feasible when the solar pressure perturbation is relatively small, and hence the method may cease to be useful for later generation, high–power, satellites with very large solar arrays.

6.2.4 Error Analysis

This is an important aspect of station–keeping analysis, and results in the derivation of the amount of the allowed deadband that should be reserved for manoeuvre implementation and tracking errors.

The errors can be divided into three groups, – execution errors, tracking errors, and modelling errors.

6.2.4.1 Execution Errors

The execution errors consist of thrust magnitude errors, thrust direction errors, and burn coupling errors.

Every time a thruster is fired there is a difference between the actual and planned total impulse delivered. The error can be divided into a bias and a random component. The former can be estimated after a few firings through orbit reconstitution on the ground. Manoeuvre planning can then take the bias into account. The random errors are more difficult to assess in orbit and ground tests must be resorted to. The effects of random errors must be allowed for in the deadband allocation.

The thrust direction is also usually non–nominal. The thrust direction error arises from several sources, i.e. thruster mounting error, thrust direction error, and attitude control error. The first can be considered to be a bias on any one satellite, while the other two will have both bias and random components. It has been found, however, that with the alignment and attitude control accuracies that can be achieved, the thrust vector misalignment is usually 0.5° at 3σ. The error that a misalignment of this magnitude causes can generally be ignored.

The error due to the coupling between the East–West and North–South station-keeping manoeuvres is important, and must be assessed and allowed for. As the North–South manoeuvre is much greater than the East–West one, when North–South station–keeping is carried out, it is the effect of the former on the latter that is of significance. This is discussed in the section dealing with North–South station-keeping, Section 6.3.3.1.

6.2.4.2 Tracking Errors

Due to the uncertainties in the orbit reconstitution and orbit evolution procedures, errors arise in the prediction of the satellite's position, and the orbit parameters.

An error in the knowledge of the satellite's *longitude* influences the timing of the correction manoeuvre and may result in the satellite over–shooting the deadband unless due allowance is made.

An error in the orbit's *semi–major axis* results in faulty planning of the manoeuvre, and the resultant drift rate could cause the satellite to exceed its dead-band or adversely influence the cycle time.

Errors in *eccentricity* and *perigee position* would affect both manoeuvre planning and the timing of the manoeuvre. Both the satellite libration and cycle time could go out of bounds unless the effects of the errors are allowed for.

Errors in the satellite *true anomaly* affect the timing of the manoeuvre.

The need to keep these errors small leads to requirements on the accuracy achieved by the tracking and orbit determination procedures. The magnitude of these requirements is mission–specific, depending upon station location, and station-keeping deadband and strategy.

Tracking errors are effectively allocated a portion of the overall deadband. For the ECS spacecraft, stationed at 10°E, and maintaining a station–keeping deadzone of $\pm 0.1^{\circ}$ in both latitude and longitude, an error analysis led to the following tracking accuracy requirements at the 1σ level.

Error in Longitude	$\leqslant 0.0065^{\circ}$
Error in Semi-Major Axis	$\leqslant 0.07$ km
Error in Eccentricity	$\leqslant 2\%$
Error in Right Ascension of perigee	$\leqslant 1^{\circ}$
Error in Mean Anomaly	$\leqslant 2^{\circ}$

These are typical figures and are presented here to give some idea of the tracking accuracy required.

6.2.4.3 Modelling Errors

During the discussion of the perturbation mechanism, second order effects were ignored. These do, however, cause some variation about the mean, and should be taken into account by the allocation of a portion of the deadband to all effects not accurately modelled in the station-keeping analysis.

6.2.4.4 Synthesis of Errors

Once the influence of each error source has been analysed, their combined effect must be assessed. In general, this is done by adding the mean (or bias) components and root-sum-squaring (RSS-ing) the random components to arrive at the deadband allocation for errors. The way the errors are synthesised also depends upon the manoeuvre strategy used.

The manoeuvre strategy is designed to minimise propellant useage. This implies that tracking errors and manoeuvre execution errors must be allowed for when allocating the available East-West deadband. One such strategy is outlined below. It allows for 99% probable errors, and hence errors to the level of 2.58 standard deviations (2.58σ) are considered.

The available deadband consists of four basic sub-divisions as follows:

- uncontrolled second-order perturbations;
- longitude excursion due to secular acceleration;
- libration due to eccentricity;
- tracking and execution errors.

A band of 0.0075° is reserved at each end of the total deadband to accommodate uncontrolled second-order perturbations. This comprises 0.0025° for the twice-daily longitude perturbation caused by the Moon, and 0.005° for the fortnightly lunar perturbation.

The remaining deadband is assigned as follows – see Figure 6.13. E′ is allowed for the longitude libration resulting from eccentricity vector evolution over the duration of a maximum cycle. E is allowed for the longitude libration resulting from eccentricity vector evolution over half a maximum cycle. B is the root-sum-squared value of the longitude tracking error and the libration due to errors in the eccentricity control manoeuvre. X indicates the *switching boundary*, i.e. the station-keeping manoeuvre is initiated when the satellite's measured position reaches this value. V and U are obtained by stepping on either side of X by an amount C which is equal to $(1/\sqrt{2})$ times the longitude tracking error. The boundaries U and V are used to determine the maximum and minimum triaxiality cycles.

Taking V as the initial longitude, the station-keeping manoeuvre must be designed such that even with a 2.58σ error in drift-rate estimation and a 2.58σ error in manoeuvre execution compounding each other, the satellite does not overshoot the longitude limit P (which is indented from Q by an amount D which is the longitude libration due to eccentricity manoeuvre error). The maximum cycle time will occur when the satellite's initial position is at V, 2.58σ errors conspire to give a larger initial drift rate than nominal, and the final position is at U. The minimum cycle time occurs when the initial position is at U, the errors conspire to give a lower drift rate than nominal, and the satellite's final position is at V. Note that both at the start and end of the cycle the satellite's measured position is at X. This is why the value of C is $(1/\sqrt{2})$ times the longitude tracking error. The maximum and minimum cycles are illustrated in Figure 6.13.

The derivation of the allocations within a specified deadband is an iterative process, the starting point being nominal cycle duration or minimum cycle duration.

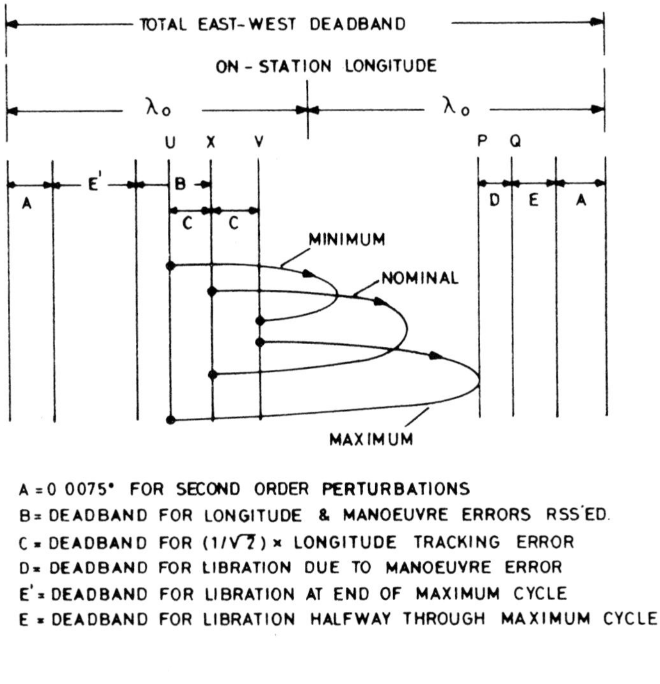

Fig.6.13 East-West Station-Keeping Strategy

6.3 North-South Station-Keeping

A geostationary satellite drifts in latitude due to the influence of the gravitational attraction of the Sun and the Moon. The perturbations caused by the Sun and the Moon are predominantly out-of-plane effects causing a change in the inclination and in the right ascension of the orbit's ascending node. In-plane perturbations also occur, but these are second order effects and need only be considered when extremely tight tolerances, i.e. about $\pm 0.03^{\circ}$, are required. When the synchronous orbit develops a non-zero inclination, the satellite drifts in latitude as viewed by an Earth-bound observer, with a period equal to the orbit period. The maximum excursion in latitude is equal to the orbit inclination. Since it is the satellite's latitude that is visibly affected, station-keeping to counteract the luni-solar perturbations is known as North-South station-keeping.

6.3.1 The Physical Mechanism

The physical mechanism of luni-solar perturbation is similar, in general terms, to that of the Earth's oblateness. The latter causes a regression of the line of nodes, and the orbit pole, which is the normal to the orbit plane, describes a circle centred on the Earth's pole, i.e. the orbit inclination does not change. Luni-solar gravity does the same thing with respect to the ecliptic, as far as long period effects are concerned. Figure 6.14 shows the motion of the orbit pole due to the effect of oblateness, Circle A, and luni-solar perturbation, Circle B. Now, as both influences act simultaneously, the orbit pole in fact rotates about an intermediate point in approximately a circle, Circle C.

At synchronous altitude, the centre of Circle C is about 7.5° from the polar axis. If the initial condition is a geostationary equatorial orbit, and no station-keeping manoeuvres are performed, the orbit inclination will increase to a maximum of 15° after about 27.5 years, and return to zero after 55 years. Meanwhile the node line moves through 180°. The exact motion is somewhat complicated because it must be analysed on a spherical surface rather than upon a plane one, and because the Moon's inclination changes relative to the ecliptic.

For station-keeping, however, we are concerned only with pole motion in the near vicinity of the Earth's pole, and the planar concept will be useful. We shall study orbit behaviour in terms of its pole motion. We introduce the concept of the *perturbation pole*, which is defined as being the actual instantaneous direction of the perturbing body – in our case the Sun and the Moon. For oblateness perturbations, the perturbation pole is coincident with the Earth's polar axis.

6.3.2 Orbit Mechanics

As we have drawn the analogy between the effects of oblateness and luni-solar perturbations, we shall start with the formula for oblateness:

$$\dot{\Omega} = -\frac{3 J_2 \, n \cos i}{2 l^2}$$

where: $\dot{\Omega}$ = rate of change of node location,
J_2 = is the second zonal harmonic of the Earth's Potential Field,
i = orbit inclination,
l = orbit semi–latus rectum expressed in Earth radii
n = mean motion of the satellite.

This equation can be cast into a form which indicates pole motion. Since inclination is invariant, the orbit pole moves in a circle of radius (sin i). If we consider one precession period, Ω moves through 2π and the pole moves through $2\pi \sin i$.

$$\therefore \quad \dot{\Omega} = \Delta p / \sin i$$

where: Δp = angular motion of the orbit pole.

$$\therefore \quad \Delta p = -\frac{3J_2 \; n \cos i \sin i}{2l^2}$$

Note that the direction of Δp is always perpendicular to the 'perturbation' pole. It is found that the corresponding expression for luni–solar perturbations is:

$$\Delta p = \frac{3}{2}\left(\frac{m}{m_e}\right)\left(\frac{r_e}{r}\right)^3 a^3 \; n \,(1 + 2\,e^2 + ...) \cos \gamma \sin \gamma$$

where: m and m_e are the mass of perturbing body and the Earth, respectively;

r and r_e are the distance of the perturbing body and the Earth's radius, respectively;

a is the semi–major axis expressed in Earth radii;

e is the eccentricity;

γ is the angle between the orbit pole and the direction of the perturbing body (the perturbation pole).

For a circular orbit this equation simplifies to:

$$\Delta p = k_{m,s} \cdot n \;.\; \cos \gamma \;.\; \sin \gamma \tag{1}$$

with

$$k_{m,s} = \frac{3}{2}\left(\frac{m_{m,s}}{m_e}\right)\left(\frac{r_e}{r_{m,s}}\right)^3 a^3$$

where subscripts m and s refer to Moon and Sun, respectively.

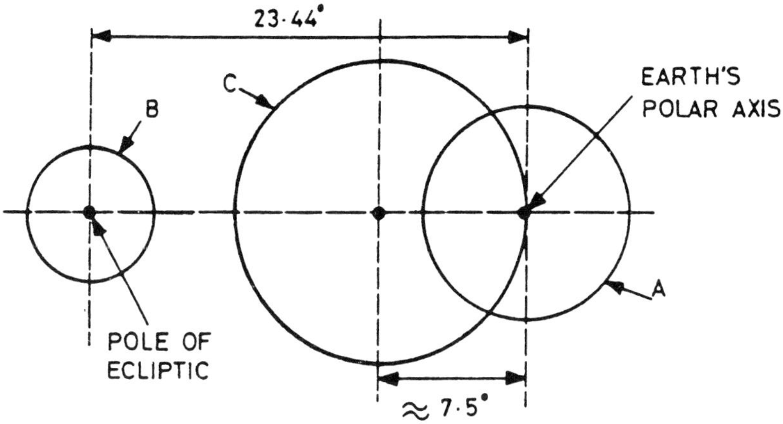

Fig.6.14 Orbit Pole Motion

Hence, the annual rate of motion of the orbit pole becomes:

$$\left.\begin{aligned} \Delta p_m &= 3.2104 \cos \gamma_m \,.\, \sin \gamma_m \\ \Delta p_s &= 1.4743 \cos \gamma_s \,.\, \sin \gamma_s \end{aligned}\right\} \text{degrees/year} \qquad (2)$$

Equation (1) is strictly valid only if the distance of the perturbing body is very large compared with the distance of the perturbed body. For the Moon perturbing a synchronous satellite, this ratio is about 9.1 and hence the perturbing effects of the Moon are about 2% larger than shown above. With this correction, Equation (2) becomes:

$$\left.\begin{aligned} \Delta p_m &= 3.275 \,.\, \cos \gamma_m \,.\, \sin \gamma_m \\ \Delta p_s &= 1.475 \,.\, \cos \gamma_s \,.\, \sin \gamma_s \end{aligned}\right\} \text{degrees/year} \qquad (3)$$

The above expressions can also be written in a form that indicates the velocity change required to counteract the pole motion. One degree of pole motion can be counteracted by a velocity change given by:

$$\Delta V = 2\, V_c \sin (½)$$

where V_c = velocity of satellite in synchronous orbit
= 3075 m/s.

∴ $\Delta V = 53.67$ m/s

Hence Equations (3) can be rewritten:

$$\left.\begin{aligned} \Delta V_m &= 175.76 \cos\gamma_m \;.\; \sin\gamma_m \\ \Delta V_s &= 79.16 \cos\gamma_s \;.\; \sin\gamma_s \end{aligned}\right\} \text{(m/s)/year} \tag{4}$$

To derive the pole motion we must evaluate the variation of γ_m and γ_s with time. Refer to Figure 6.15. The Sun is projected on a sphere of unit radius. The co-ordinates of the projection are:

$$\begin{aligned} x &= \cos\beta \\ y &= \sin\beta\cos\epsilon \\ z &= \sin\beta\sin\epsilon \\ x^2 + y^2 + z^2 &= 1 \end{aligned}$$

where β = position of Sun in ecliptic ($\beta = 0$ at vernal equinox)
ϵ = angle between Equator and ecliptic.

Now if we generalise Equations (3) and (4) so that Δ stands for Δp or ΔV and k for the appropriate constant, we can write:

$$\Delta = k \sin\gamma\cos\gamma = k \;\; z\sqrt{1-z^2}$$

because from Figure 6.16:

$$\sin\theta = \cos\gamma = \sin\epsilon\sin\beta = z$$

and $\sin\gamma = (1 - z^2)^{1/2}$

Hence $\Delta = k \;.\; z\sqrt{x^2 + y^2}.$

Resolving into the x and y components of Δ we have:

$$\Delta_x = \Delta \frac{y}{\sqrt{x^2 + y^2}} = k \,.\, z \,.\, y$$

$$\Delta_y = -\Delta \frac{x}{\sqrt{x^2 + y^2}} = -k \,.\, z \,.\, x$$

The direction of $\dot{\Delta}$ is perpendicular to the perturbation pole.

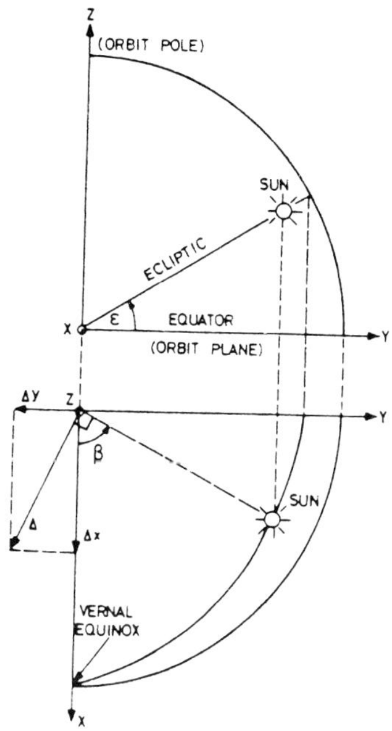

Fig.6.15 Luni–Solar Perturbation Geometry

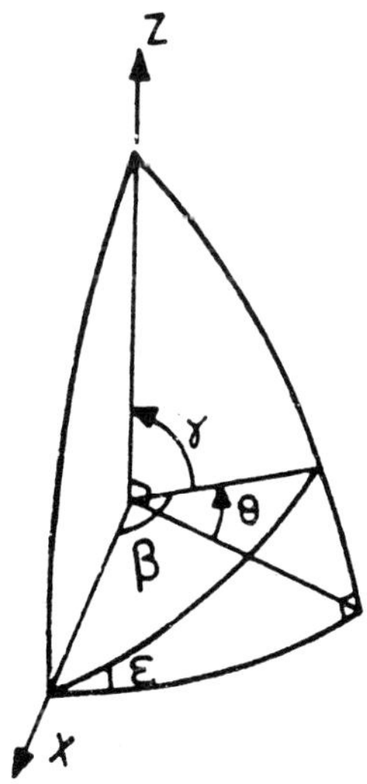

Fig.6.16 Spherical Triangles Showing ϵ, β, θ and γ

$$\Delta_x = k \sin\epsilon \cos\epsilon \sin^2\beta$$

$$= \frac{k}{2} \sin\epsilon \cos\epsilon\,(1 - \cos 2\beta)$$

$$\Delta_y = -k \sin\epsilon \sin\beta \cos\beta$$

$$= \frac{-k}{2} \sin\epsilon \sin 2\beta$$

Now, integration over β yields:

$$\int \Delta_x \, d\beta = \frac{k}{2} \sin\epsilon \cos\epsilon \left[\beta - \frac{\sin 2\beta}{2}\right]$$

$$\int \Delta_y \, d\beta = \frac{k}{2} \sin\epsilon \frac{\cos 2\beta}{2} \tag{5}$$

Equations (5) indicate that the motion of the pole consists of a secular motion Δ_s in the direction of the positive X–axis at an annual rate of:

$$\Delta_s = \frac{k}{2} \sin\epsilon \,.\, \cos\epsilon$$

and a periodic motion, with a period of half a year, in an ellipse with amplitudes Δ_{px} and Δ_{py} in the directions of the X and Y axis respectively of:

$$\Delta_{px} = \frac{k}{2} \sin\epsilon \cos\epsilon \;.\; \frac{1}{4\pi}$$

$$\Delta_{py} = \frac{k}{2} \sin\epsilon \,.\, \frac{1}{4\pi} \tag{6}$$

For the Sun, using the k - value of 1.475° we get:

$$\Delta_s = 0.2691^\circ$$
$$\Delta_{px} = 0.0214^\circ$$
$$\Delta_{py} = 0.0233^\circ$$

The analysis shown above for solar perturbations can also be repeated for lunar perturbations bearing in mind that the inclination of the Moon's orbit is at 5.135° to the ecliptic and hence ϵ varies between 28.58° and 18.31°. The maximum occurred on 27 March 1969, the minimum on 19 July 1978. In addition, the direction of the ascending node of the Moon's orbit can deviate as much as 13° from the Vernal Equinox, thus affecting the direction of secular motion. A further difference is that the constant k in equations (6) must now refer to a rate based upon the sidereal period of the Moon, 27.32 days. Thus the results of equations (6) must be divided by 365.25/27.32 = 13.37. Using a k -value for the Moon of 3.275° we have:

		1969	1978
Δ_s	=	0.6878°	0.4884°
Δ_{px}	=	0.0041	0.0029
Δ_{py}	=	0.0047	0.0031

It can be seen that though the secular motion due to the Moon is much larger than that due to the Sun, the periodic motion is so small that it can generally be ignored.

If we now use the k – values from Equation (6) we can obtain the ΔV required to compensate for the pole motion. This can be done for the secular component Δ_s, and yields:

	1969	1978	
ΔV_{sun}	14.45	14.45	(m/s)/year
ΔV_{moon}	36.93	26.35	(m/s)/year
ΔV_{total}	51.38	40.80	(m/s)/year

The variation of total ΔV with year is shown in Figure 6.17. This indicates the ΔV required during only one year to compensate for secular pole motion. It must be noted that in the above, we use the term secular rather loosely. The motion is in reality of the long period type; only in the context of station–keeping, when the orbit pole position is maintained within the close proximity of the Earth's pole, can this motion be thought of as secular.

6.3.3 The Manoeuvre Strategies

The North–South station–keeping requirement for a large class of geosynchronous satellites is about $\pm 0.1^{\circ}$ in latitude about the Equator. For such a deadband it is necessary only to compensate for the *secular* drift due to luni–solar perturbations. In this case, for a specific mission, the minimum ΔV required can be obtained from Figure 6.17, where it is assumed that the manoeuvres are applied at the optimum time of day. This optimum time occurs when the direction of the satellite from the centre of the Earth is perpendicular to the desired direction of motion of the orbit pole.

As the direction of secular pole motion due to lunar perturbations depends upon the time being considered, the manoeuvre times will also vary. For station–keeping purposes, the general direction of orbit pole motion due to perturbations is in the direction of the Vernal Equinox. The optimum manoeuvre positions will therefore vary about the 90° and 270° positions relative to the Vernal Equinox, the exact positions being time dependent.

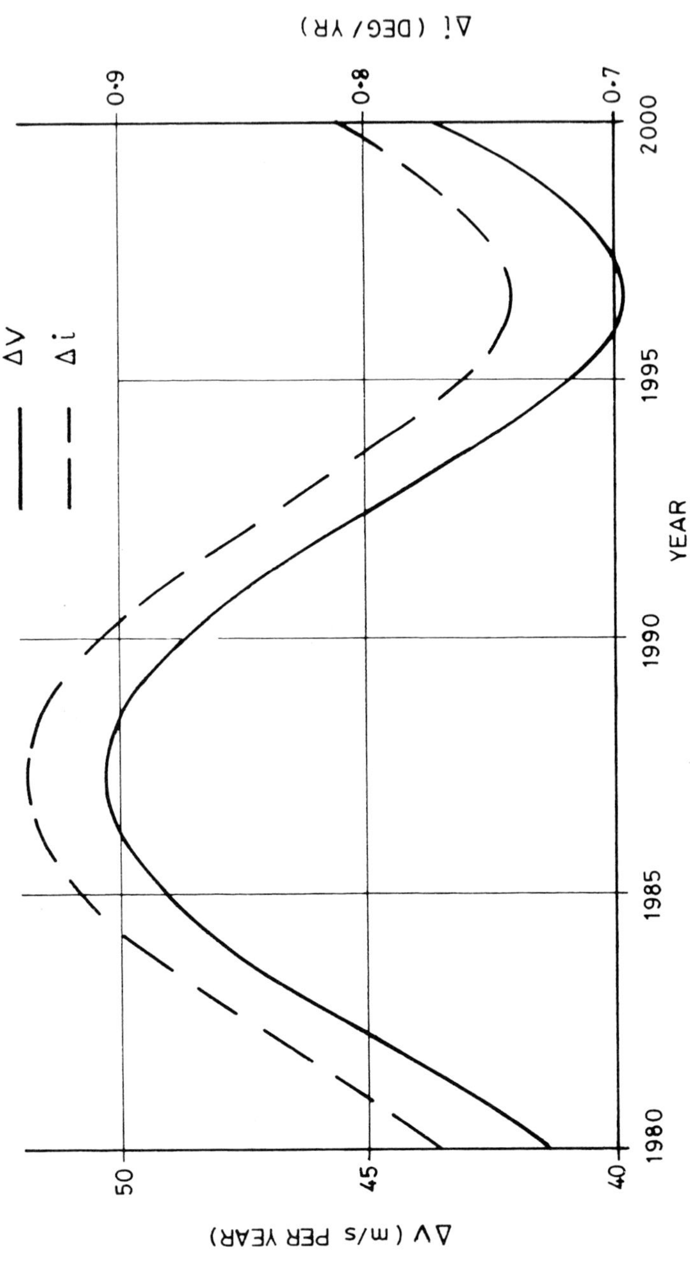

Fig.6.17 Annual ΔV for North-South Station-Keeping

Sometimes it may be necessary to perform the manoeuvres non–optimally. In such a case, referring to Figure 6.18, where C is the position of the optimum manoeuvre, the burns can be performed at A and B instead of at C, and the impulses at A and B are inversely proportional to the distances AC^1 and BC^1, the total impulse being increased over the nominal one in the ratio CM/CM^1.

As indicated by the ΔV required to compensate for luni–solar perturbations, the burn magnitudes required for North–South station–keeping are much greater than those for East–West station–keeping. Any misalignment of the thrust direction away from nominal will therefore result in a non–negligible thrust component in the East–West direction. This will have to be counteracted by the East–West station-keeping manoeuvre. To enable this to be done efficiently, the North–South manoeuvre should be arranged to take place sufficiently before the end of the East-West station–keeping cycle. The effect of the North–South burn can then be assessed, and the required compensation carried out during the East West burn. It should be noted that the North–South burn errors will have both a mean and a random component. The random component can be allowed for in the East–West station-keeping strategy by the reservation of a certain proportion of the available dead-band for these errors. If this is done, there is no ΔV penalty to be incurred. The mean component is, however, an effective change in the satellite's mean drift rate, and a ΔV allowance must be made to counteract this effect. It should be realised that the ΔV allowance for this could be of the same order as the nominal East-West station–keeping budget. In the section below we shall outline the North–South station–keeping strategy adopted for ESA's OLYMPUS spacecraft.

6.3.3.1 "OLYMPUS" North–South Station–Keeping

The North–South station–keeping strategy to be adopted for ESA's OLYMPUS spacecraft has been developed at the European Space Operations Centre (ESOC) and is used also for the ECS family of spacecraft. Here we shall give an outline of its main elements.

The North–South station keeping strategy corrects only for the secular motion of the orbit pole, but by appropriate targeting, minimises the North–South libration of the spacecraft and so results in a most efficient use of station–keeping propellant. The strategy consists of setting up the initial conditions of the orbit pole in such a way that the orbit inclination at the start and end of the cycle is the same.

Let us define the position of the orbit pole in the X–Y plane lying in the Equatorial plane, with the X–axis directed towards the Vernal Equinox. From Figure 6.19 it can be seen that the x and y co–ordinates of the orbit pole are given by:

$$x = i \sin \Omega \quad ; \quad y = -i \cos \Omega$$

where: i = orbit inclination,
Ω = right ascension of the ascending node.

Since we know that the motion of the orbit pole comprises secular and periodic motions, let us express each co–ordinate in those terms. Therefore:

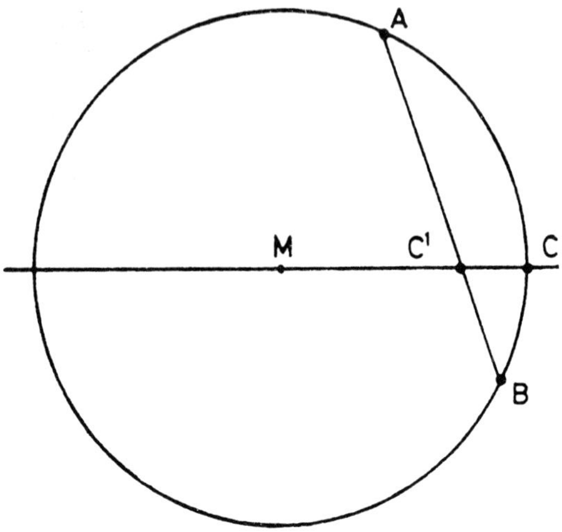

Fig.6.18 Manoeuvre Efficiency

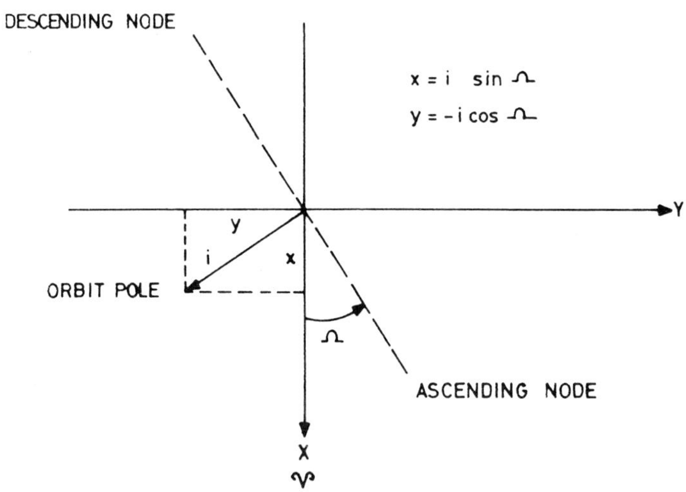

Fig.6.19 Definition of Orbit Pole

$$x = x_s + x_p \quad ; \quad y = y_s + y_p$$

where: subscript s refers to secular motion, and subscript p refers to periodic motion.

We now define the following:

t_o = station–keeping cycle start time,

T = station–keeping cycle duration,

$[x_o(t_o);y_o(t_o)]$ = instantaneous pole position at start of cycle,

$[x_e(t_o+T),y_e(t_o+T)]$ = instantaneous pole position at end of cycle,

$(\dot{x}_s, \dot{y}_s)$ = rate of change of pole position due to secular effects,

$[x_p(t),y_p(t)]$ = offsets from secular motion at time t due to periodic effects.

As the initial and final inclinations are the same, we have:

$$x_o(t_o)^2 + y_o(t_o)^2 = x_e(t_o+T)^2 + y_e(t_o+T)^2 \tag{1}$$

Let us initialise the cycle by setting the y–co–ordinate of the secular motion to zero. Thus we have:

$$y_o(t_o) = y_p(t_o) \tag{2}$$

The y–co–ordinate of the instantaneous pole motion at the end of the cycle is, therefore:

$$y_e(t_o+T) = \dot{y}_s T + y_p(t_o+T)$$

The x–co–ordinate of the instantaneous pole motion at the end of the cycle will be:

$$x_e(t_o+T) = x_o(t_o) - x_p(t_o) + \dot{x}_s T + x_p(t_o+T) \tag{3}$$

The first two terms on the right hand side of the Eqn.(3) give the secular motion at the start of the cycle. To this is added the secular motion in the x–direction during the duration of the cycle to give the x–co–ordinate of the secular motion at the end of the cycle. Finally, the periodic offset at the end of the cycle is added.

We now solve for the x–co–ordinate at the start of the cycle and obtain:

$$x_o(t_o) = -\tfrac{1}{2}\ [x_p(t_o + T) - x_p(t_o) + \dot{x}_s T] \quad + \quad B$$

where:

$$B = \frac{y_p(t_o)^2 - y_p(t_o + T)^2 - (\dot{y}_s T)^2 - 2\dot{y}_s T\, y_p(t_o + T)}{2\ [x_p(t_o + T) - x_p(t_o) + \dot{x}_s T]} \qquad (4)$$

It can be seen that the start position depends upon the pole position at the end of the forth-coming cycle, and the periodic motion components at the start of the cycle. Equations (2) and (4) define the pole position at the start of the cycle. They are solved by evaluating the effect of luni-solar perturbations through orbit propagation. The pole position will evolve from $[x_o\ (t_o),\ y_o\ (t_o)]$ to $[x_e\ (t_o + T);\ y_e\ (t_o + T)]$ over the time interval T, at which time Eqn.(1) will apply. A North-South station-keeping manoeuvre would then be designed to move the pole from that point to $[x_o\ (t_o + T);\ y_o\ (t_o + T)]$. The magnitude of the manoeuvre is given by:

$$\Delta V = V_c\ [(x_o(t_o + T) - x_e(t_o + T))^2 + (y_o(t_o + T) - y_e(t_o + T))^2]^{\frac{1}{2}}$$

where: V_c = velocity of the spacecraft in geosynchronous orbit.

The right ascension at which the burn is applied is given by:

$$\Theta_N = \tan^{-1}\left[\frac{x_o(t_o + T) - x_e(t_o + T)}{y_o(t_o + T) - y_e(t_o + T)}\right]$$

for northerly thrusts, and

$$\Theta_s = \Theta_N + \pi$$

for southerly thrusts.

Neither the magnitude nor the position of the manoeuvre will remain constant with time. The spacecraft will, however be maintained within the minimum inclination limits for any given cycle duration, for a minimum expenditure of propellant.

6.3.3.2 The Free-Drift Strategy

One way of increasing the useful payload that a satellite can carry is to reduce the propellant requirement for the on-station phase of the mission. Since most of the propellant in this phase is used for North-South station-keeping, a considerable reduction in propellant can be achieved by eliminating, or greatly reducing, this activity. Abandonment of North-South station-keeping does, however, result in the spacecraft moving in latitude (and to a very small extent in longitude also) with respect to an Earth-fixed observer. Some communications missions, particularly those employing very wide coverage beams, can tolerate such motion, and for these abandonment of North–South station–keeping becomes an attractive option. For example, the MARECS satellite is allowed a North–South deadband of $\pm 3.0^{\circ}$ in latitude for its 7–year lifetime.

Even when North–South station–keeping is abandoned, it is important to minimise the satellite's orbit inclination as far as possible. This minimises the satellite's latitude excursions and increases the efficiency of communications. Since the orbit pole, if uncontrolled, moves in the direction of the Vernal Equinox, placing it initially on the side remote from the Vernal Equinox results in the orbit inclination first reducing, then passing through a minimum before increasing again. The motion of the orbit pole is illustrated on the polar diagram of Figure 6.20 where the inclination is plotted radially while the right ascension of the ascending node is plotted circumferentially. There is a unique value of pole position which maximises a satellite's lifetime within a specified inclination limit. If the satellite's design lifetime is less than the maximum achievable within the specified inclination limit, a range of initial pole positions becomes available. The range of initial pole positions is shown in Figure 6.20 and is defined by the arcs PAQ and PBQ. The boundary PBQ defines the maximum inclination limit, while the arc PAQ is defined by the initial pole positions which give exactly the design lifetime below the specified inclination limit. On Figure 6.20, the region bounded by the arcs RB'S and RA'S corresponds to the pole position at the end of the design life. The arc PBQ maps on to arc RB'S and RAQ on to RA'S. The lines AA' and BB' are two typical traces of pole evolution through the lifetime.

If the maximum dwell time during free drift within the specified inclination limit is less than the satellite's design lifetime, pole position control becomes necessary. The optimum pole position can now be defined as that which results in the minimum propellant useage for North–South station–keeping. If, however, the propellant allowed is greater than this minimum, a range of initial pole positions will become available. One boundary of this range will still be defined by the maximum inclination limit, while the other will be defined by those initial pole positions which would require the full allocation of North–South station–keeping propellant. For pole positions inside this region, the propellant used for North–South station–keeping to achieve the desired lifetime will be less than that allocated.

Several strategies are available for the implementation of pole position control. These are:

(1) a free–drift strategy with North–South station–keeping within tight inclination limits, say $\pm 0.1^{\circ}$, about the minimum inclination;

(2) a free–drift strategy which takes the orbit pole, across the maximum inclination circle, to the inclination limit, followed by a manoeuvre to set up the correct pole position to allow free drift for the satellite's remaining lifetime;

(3) a free–drift strategy, followed by a manoeuvre early in the lifetime to restore the inclination to the maximum, allowing the remainder of the lifetime to be free of North–South manoeuvres;

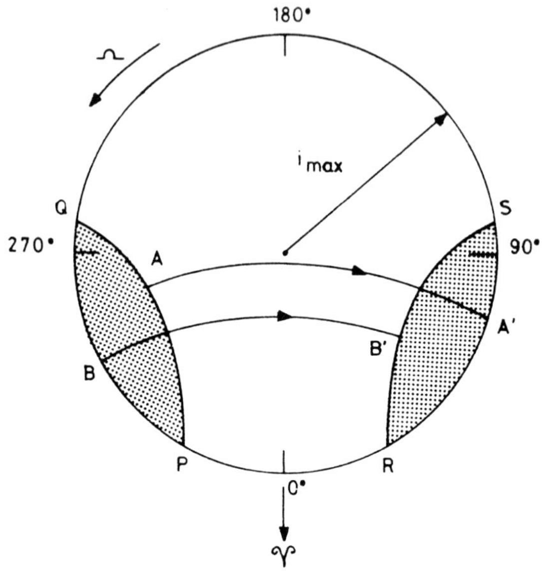

Fig.6.20 Pole Motion for Free-Drift Strategy

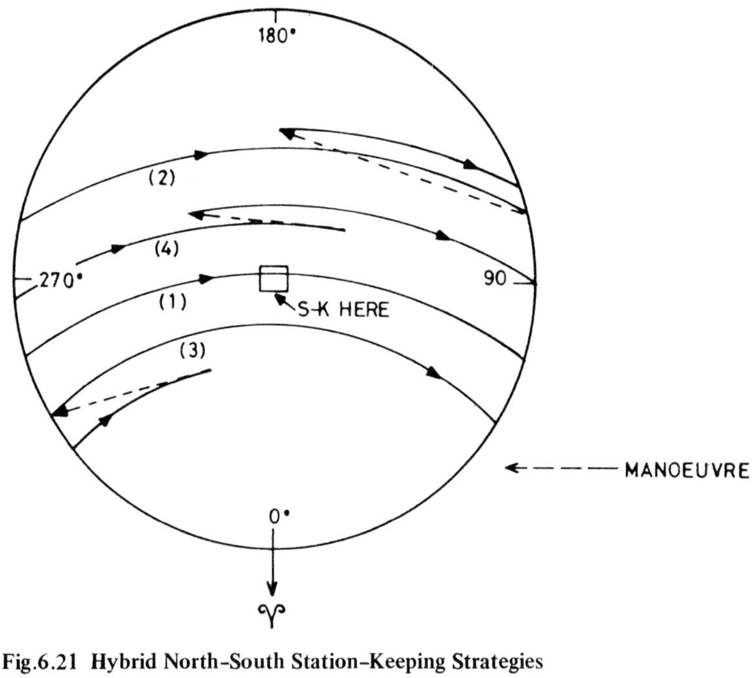

Fig.6.21 Hybrid North-South Station-Keeping Strategies

(4) essentially the same as (2) and (3) except that the manoeuvre is made in such a way as to allow the satellite more time at low inclinations than at high;

(5) a free-drift strategy with small corrections applied regularly through the lifetime to slow down the rate at which the orbit pole moves across the maximum inclination circle.

These strategies are illustrated in Figure 6.21. Strategies (2) and (3) require large changes in pole position. This implies large manoeuvres, and with the size of thrusters normally used for station-keeping it will not be possible to execute the manoeuvre as a single burn. A series of burns, over several days, will be required. The manoeuvre planning must, therefore, consider the pole motion drift while the manoeuvre campaign is in the process of implementation. In addition, the manoeuvres may cause perturbations in the East-West direction and cause some difficulties in maintaining the East-West deadband. Strategy (5) alleviates this problem by ensuring that the North-South manoeuvres are small. By careful planning, a North-South manoeuvre may be executed in such a way as to allow any perturbation in the East-West direction to be measured and evaluated. It can then be accounted for in the planning of the next East-West station-keeping manoeuvre.

6.3.4 Error Analysis

In Section 6.2.4.1 we considered the influence of manoeuvre errors upon the East-West station-keeping strategy. In this section we shall consider the impact of manoeuvre errors upon the North-South station-keeping fuel budget.

If the thrusters are misaligned then a direct fuel penalty will result. However, thruster alignment is usually closely controlled and the thrust vector direction is usually accurate to within 0.5^{o}. The cosine loss due to this misalignment is 0.004% and can be ignored.

A further source of error is in the delivered thrust level. This can be divided into mean and random components. The mean component can usually be derived after a few manoeuvres through orbit reconstitution. Some idea of thrust level will generally be obtained during the station-acquisition and initialisation manoeuvres. The mean error, once known, can be allowed for when designing subsequent manoeuvres, and should, therefore, impose no significant penalty. The random error can be assessed through ground tests and a portion of the North-South deadband must be reserved for the orbit pole motion resulting from a variation of thrust level. This will lead to a variable North-South station-keeping cycle time but will result in zero fuel penalty.

It can be seen, therefore, that the fuel penalty due to manoeuvre implementation errors can be considered to be essentially zero. This is fortunate, as the fuel required for North-South station-keeping is very large, and a further penalty would be most inconvenient.

In addition to thruster errors, we also have to allow for tracking errors, and this is done by allocating a further portion of the deadband to allow for pole-prediction errors. Once again this results in a variation of cycle time, but leads to no fuel penalty.

6.3.5 Plume Impingement

When evaluating the propellant required to implement a station-keeping strategy, the efficiency of the manoeuvre must be taken into account fully. The efficiency of the manoeuvre is influenced by two factors. The factor related to the position in the orbit has been referred to in Section 6.3.3. The other factor comes about due to the effects of plume impingement. When a thruster fires, the efflux – also known as the plume – may impinge on parts of the spacecraft and partially negate the effect of the burn. Spacecraft which are 3-axis stabilised when on station generally have solar panels deployed along the North–South axis. There is, therefore, ample scope for adverse interaction between the solar panels and the thruster plumes during periods of North–South station-keeping. Plume impingement has two undesireable effects. First, it produces a *drag* effect directly opposing the direction of the thrust, and thus reducing its effectiveness. Second, it produces disturbance torques that must be counter-acted if the pointing accuracy of the spacecraft body is to be maintained. The extent of plume impingement depends on the thruster and solar array geometry, and is therefore time dependent, with a period of one year. Integrating the curve of *drag* loss against time over the period of one year yields the annual *drag* loss for the North–South station-keeping activity. This tells the mission designer how much extra propellant he must carry on the spacecraft to offset the *drag* effect. Usually plume losses can be reduced by adjusting the thrust line of the thrusters. The thruster orientation is selected through an optimisation procedure, leading to minimum use of station-keeping propellant, by balancing plume losses and cosine losses. For ESA's OLYMPUS spacecraft, the average loss of efficiency due to a combination of plume and cosine losses is about 13%. For the EUROSTAR spacecraft the corresponding figure is about 6%.

6.4 The Station-Keeping of Spacecraft Clusters

In this section we shall consider the ways in which station-keeping can be implemented for spacecraft clusters. Before we do this, however, a short discussion on spacecraft clusters will set the scene for the following treatment.

The growth trend in telecommunications is such that it will soon exceed the capability of present and planned launch vehicles to satisfy payload requirements. One solution to the problem is to divide the payload into several segments, mount each segment on a separate spacecraft, and operate the collection of spacecraft in such a way that they appear as one to the Earth-based payload user. This is the concept of the satellite cluster, wherein several separate and distinct satellites co-operate to provide a level of service larger than each can provide on its own. The size and characteristics of the cluster would depend upon the nature of the payload. Several variations are possible, ranging from master/slave configurations to co-equal configurations, configurations with full telecommunications inter-connectivity, and those without any inter-connectivity. The cluster concept not only provides a release from the payload limitation of launch vehicles, but also allows the system to be implemented in stages to match growing demand.

Whatever the detailed nature of the cluster, the concept is only attractive if the division of the payload between several spacecraft is transparent to the user. For some configurations, this implies that all the cluster members come within the beamwidth of a single, fixed, ground station antenna. Current practice requires communications spacecraft to be kept within approximately $\pm 0.1^{\circ}$ in latitude and longitude, and this is gradually being tightened to about $\pm 0.05^{\circ}$. It follows, therefore, that the entire cluster must also satisfy these station-keeping requirements. This implies tighter control of a spacecraft's position and velocity (both in absolute and relative terms) than has been necessary so far. As the station-keeping strategy is closely dependent upon cluster geometry, we shall briefly examine some of the possible cluster geometries.

6.4.1 Cluster Geometries

Cluster geometries fall into two main classes:

Class 1: geometries in which all the satellites share the same mean longitude.

Class 2: geometries in which the satellites have small differences in their mean longitudes.

A third, hybrid, class also exists in which some, but not all the satellites have the same mean longitude.

The simplest geometry is in fact a Class 2 geometry in which the satellites are placed side-by-side along the geostationary arc. Such a configuration, however, suffers from the disadvantage that inter-satellite sight-lines are co-incident. Only near neighbours can have communication links between them. Introducing small eccentricities and inclinations into some of the satellites of a *necklace* cluster generates a series of Class 2 geometries with sight-line separation. Figure 6.22 shows such a geometry.

Larger angular separations between sight-lines are provided by Class 1 geometries. The spacecraft of the cluster are co-planar, and revolve around each other through 360° per day. Two such geometries have been explored, the more interesting of which is called the *tilted circle* geometry. The orbit characteristics of a four-satellite cluster are shown below:

$$\left.\begin{aligned} e_j &= 1.186 \times 10^{-4} \\ i_j &= 1.177 \times 10^{-2} \text{ degs} \\ \omega_j &= 270^{\circ} \end{aligned}\right\} j = 1, 2, \ldots\ldots N$$

$$\begin{aligned} \Omega_1 &= 0^{\circ} & M_1 &= 180^{\circ} \\ \Omega_2 &= 90^{\circ} & M_2 &= 90^{\circ} \\ \Omega_3 &= 180^{\circ} & M_3 &= 0^{\circ} \\ \Omega_4 &= 270^{\circ} & M_4 &= 270^{\circ} \end{aligned}$$

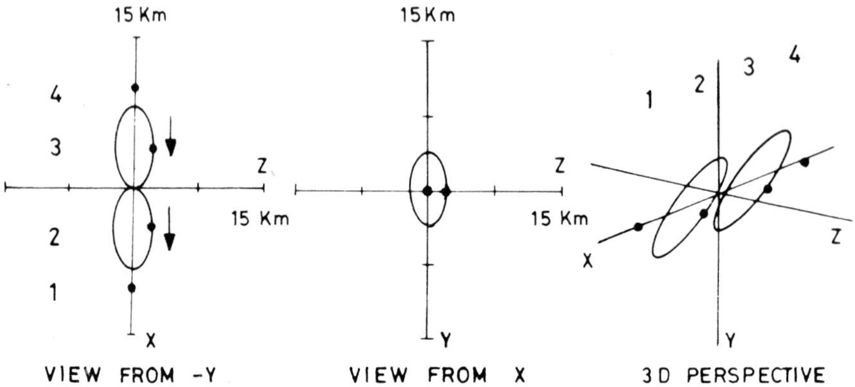

Fig.6.22 A Class 2 Satellite Cluster Geometry

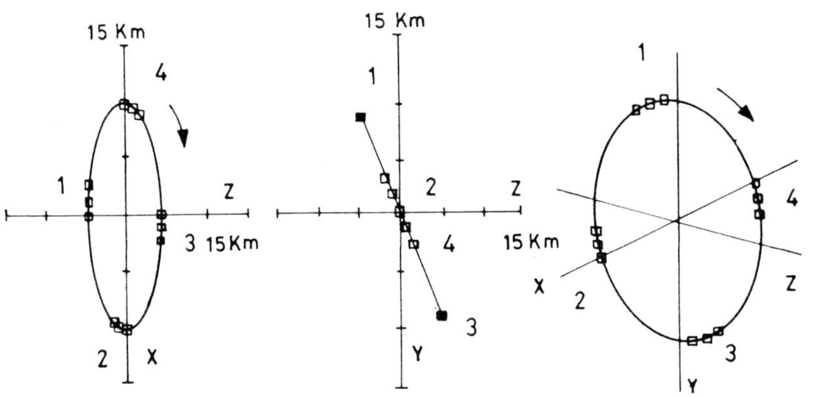

Fig.6.23 The Tilted Circle Geometry

where: e = eccentricity,
i = inclination,
ω = argument of perigee,
Ω = right ascension of the ascending node,
M = mean anomaly,
N = number of satellites in the cluster.

The cluster is illustrated in Figure 6.23. The diameter of the circle is 20 km, and the minimum sight–line separation angle is 45°.

6.4.2 Evolution of Cluster Orbits

Each satellite of the cluster is subject to the orbit perturbations which form the subject of this chapter. The effect of these perturbations and others on the cluster geometry is a matter of interest when considering the feasibility of the cluster concept. It should be noted that perturbations in the orbit elements are, after a short time, of the same order as the differences in the nominal values between satellites.

Triaxiality perturbs the semi–major axis of a satellite orbit, causing a drift in longitude. The members of a cluster, owing to their close proximity, can be considered to experience the same perturbing acceleration. They can, therefore, be expected to move as one, and the internal geometry should be preserved to very good accuracy within the station–keeping cycle times of interest. A possible exception is when a *necklace* geometry straddles an *unstable* longitude from the triaxiality perturbation point of view. Natural drift towards different *stable* points will cause a separation of the cluster members. This separation will, however, only occur very slowly and should be small in the cycle times of interest. The satellites must, however, be controlled to maintain the cluster.

Solar radiation pressure, to first order, affects only the eccentricity and argument of perigee of an orbit. The eccentricity vectors of the individual orbits have been found to behave in a special way. If we define a satellite's *relative eccentricity vector* as its eccentricity vector with respect to the eccentricity vector of the cluster centre, it has been shown that the orientation of the *relative eccentricity vectors* of the satellites remains fixed. This preserves the internal geometry of the cluster.

Luni–solar gravity perturbs the orbit pole. All the poles in the close vicinity of the Earth's pole suffer similar drifts. Therefore, the poles of the individual satellites within a cluster may be expected to suffer identical perturbations, thus preserving their relative geometry.

To confirm the above, orbit propagation has been carried out for the satellites of a *tilted circle* cluster geometry. The results are shown in Figure 6.24, where it is clear that the internal geometry of the cluster is essentially undisturbed by natural perturbations. In the above it has been assumed that all the members of the cluster have identical area–to–mass ratios.

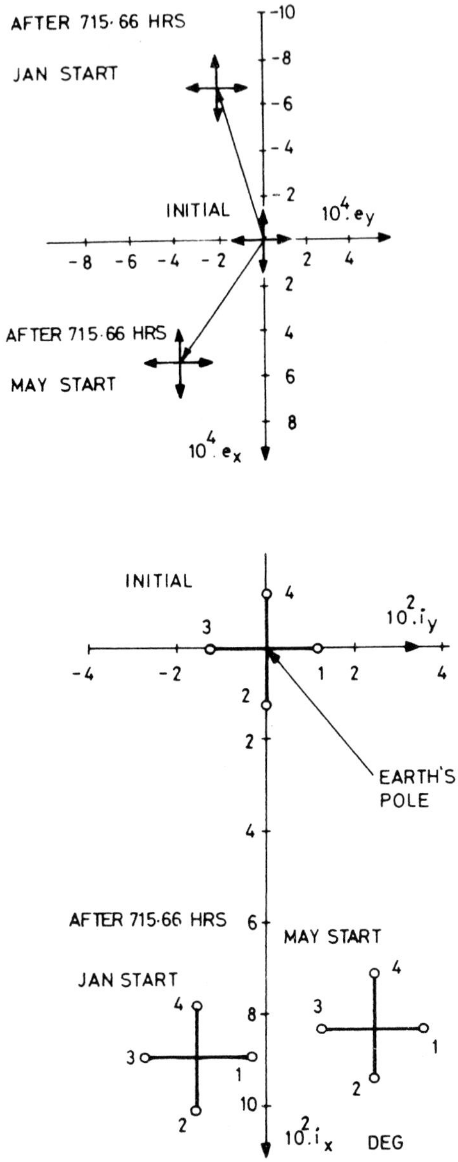

Fig.6.24 Cluster Orbit Evolution

6.4.3 Cluster Station-Keeping

If the cluster members have equal area-to-mass ratios, it has been shown that natural perturbations leave the internal geometry un-affected. An orbit control strategy may then be divided into two parts: cluster centre control, and internal geometry control.

The objective of cluster centre control is to keep the cluster as a whole within the overall deadband allocated to the mission. The cluster is considered as a single unit, and identical station-keeping burns are applied simultaneously, or nearly so, to each satellite. The design of this strategy is very similar to single satellite station-keeping described earlier in this chapter.

Although natural perturbations do not sensibly affect the internal geometry, errors in the execution of the burns will act to distort the relative configuration. Burn errors are due to random and bias errors on the thrusters, and though the latter can be progressively calibrated out, random errors will remain. A second orbit control strategy must, therefore, correct these errors if the internal geometry is to be maintained.

The orbit control of the *tilted circle* geometry is now considered in more detail. North-South station-keeping is not thought to pose a problem. However, the division of the longitude deadband for East-West station-keeping is more critical and requires further examination. A strategy is described below which is based on the control of the cluster centre, and which allows for the finite dimension of the cluster and the errors inherent in assessing and executing the strategy. The assessment is based upon the OLYMPUS spacecraft. It is assumed that a cluster of OLYMPUS spacecraft are to be maintained within an East-West dead-band of $\pm 0.1^{\circ}$ about the nominal longitude of 19° W.

The East-West deadband allocation is shown in Figure 6.25. The allocation is labelled as follows:

A : libration due to uncontrolled second-order and cyclic perturbations (0.0075°);

B : longitude tracking error (0.0073° at 3σ);

C : $1/\sqrt{2}$ of B;

D : libration due to eccentricity tracking and execution errors;

E : libration due to solar pressure perturbations (0.03°);

F : half the East-West angular extent of the cluster (0.014°);

G : spare deadband (0.01°);

S/B : the switching boundary, i.e. manoeuvres are executed when the cluster centre is estimated to be at S/B;

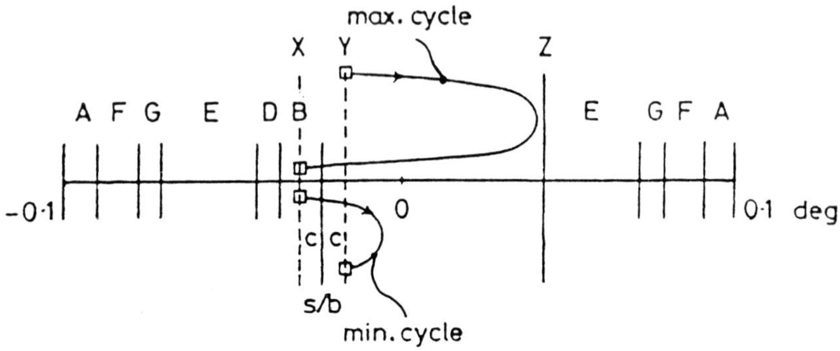

Fig.6.25 Cluster East-West Station-Keeping

X,Y : define the limits of the start and end points of the maximum and minimum triaxiality cycles;

Z : the maximum excursion due to triaxiality drift.

The mean features of the strategy are:

- the triaxiality deadband allocated to the cluster centre is 0.05^{o}. Triaxiality perturbations alone will therefore keep the cluster centre within a deadband of 0.05^{o} even allowing for tracking and execution errors at the 99% probability level;

- the deadband allocated to solar pressure perturbations of the cluster centre is 0.06^{o}. Hence the cluster centre eccentricity is such that in the face of execution errors the satellite libration is never more than 0.03^{o};

- the deadband allocated to the finite East-West extent of the cluster is 0.028^{o};

- the remaining deadband allocations are to allow for second-order and cyclic perturbations, longitude tracking error, and errors in setting up the desired eccentricity vector.

Such a strategy has a nominal cycle time of 14 days, and maximum and minimum cycle times of 17 and 10 days respectively. The annual velocity increment is about 6 m/s per spacecraft. The internal geometry will, however, be perturbed owing to burn errors. For OLYMPUS, with liquid bi-propellant thrusters, the 99% probable burn error is taken to be 2.5% of nominal. The mean error is therefore 0.56%. If internal geometry is to be maintained, these burn errors must be removed by corrective manoeuvres. The fuel penalty would therefore be about 0.56% of 6 m/s, and can be seen to be a very small amount.

Adding further spacecraft to the *tilted circle* geometry does not affect the cluster centre control strategy. This is because the distribution of the eccentricity and pole vectors does not spread with increasing member numbers. The toierance on the internal geometry may, however, be expected to become much tighter.

6.4.4 Orbit Determination

A pre-requisite for orbit control is the accurate determination of the orbits to be maintained. The table below gives a comparison between the tracking accuracy achievable using current techniques, and the accuracy demanded by cluster control.

Orbit Parameter	**Current Accuracy (1σ)**	**Cluster Requirement (1σ)**
semi-major axis (km)	0.02	0.003
eccentricity	4.6×10^{-6}	4.6×10^{-6}
inclination (deg)	0.002	0.0005
longitude (deg)	0.0055	0.00084

The values for current tracking accuracy are based upon typical measurements (i.e. range and two angles) from a single ground station with a tracking interval of 1 hour over a period of 48 hours. The accuracy demanded by clusters has been taken as 10% of the nominal differences between satellite orbit elements. The semi-major axis accuracy requirement has been based on the need to avoid collision, between spacecraft of a four-satellite necklace cluster, within a period of 10 days.

To achieve the above requirements some improvement in tracking performance is necessary. This could come in several ways:

- use of more than one ground station so located as to give good triangulation on the spacecraft;
- more accurate tracking measurements (using laser ranging, for example);
- use of new techniques involving the tracking of a *pilot* satellite, and *relative* tracking of the other satellites in the cluster.

Studies can be found in the recent literature dealing with the orbit control aspects of cluster establishment and maintenance. The station-keeping of clusters is seen to be a difficult task requiring developments of both on-board capabilities and ground station performance.

6.5 Future Trends and Requirements

A study of the evolution of spaceborne communications systems indicates that spacecraft will increase in number, increase in capacity, and be required to achieve greater reliability. The first trend will lead to a crowding of the geostationary orbit, the second to larger satellites, and the third to longer design lifetimes. All three trends will have a direct impact upon the station-keeping requirements.

Greater communications capacity will be achieved through a move to higher frequencies, and this will lead to a progressive tightening of the East-West station-keeping deadband in order to minimise ground station complexity. Tighter deadbands, for the above reason and also to accommodate more spacecraft, require greater ΔV for station-keeping and higher tracking accuracy. The former will provide the impetus for more efficient propulsion systems, while the latter could result in novel tracking techniques such as laser tracking or satellite-to-satellite tracking. One means of overcoming congestion is to use spacecraft in 24-hour orbits of non-zero inclination. This could double or treble the orbit capacity, but would require a family of spacecraft and fairly complex ground stations. The station-keeping requirements would also be quite severe, with inclination maintenance imposing up to a 10%–20% penalty in fuel compared to a zero-inclination orbit. A family of spacecraft, typically 4 to 8, could provide two *virtual* satellites at latitudes of approximately $\pm 15^{\circ}$.

The increase in communications capacity will lead to larger satellites through larger solar arrays and antennae. The trend for an increasing area-to-mass ratio is already well established. This will lead to a dominance of solar pressure perturbation over that related to the ellipticity of the Earth's equatorial cross-section, and will result in an increase in the East-West station-keeping fuel requirement. Furthermore, these large structures will be relatively more flexible than current spacecraft, and the spacecraft's control system and manoeuvre strategy will require to be carefully designed to avoid de-pointing of the antennae during station-keeping manoeuvres.

The increase in spacecraft reliability will lead to spacecraft lifetimes substantially greater than those of today. This characteristic, coupled with the requirement of station-keeping within small North-South and East-West deadbands, will point to the use of higher efficiency propulsion systems, possibly electric propulsion. This in turn will lead to manoeuvres being required every few days or so. Though this would to some extent alleviate the severity of the structure-control system interaction problem, the frequency of the manoeuvres would impose a high workload on the ground station controlling the spacecraft. The obvious answer to this problem is to make the station-keeping activity autonomous, freeing the spacecraft from ground intervention for this function.

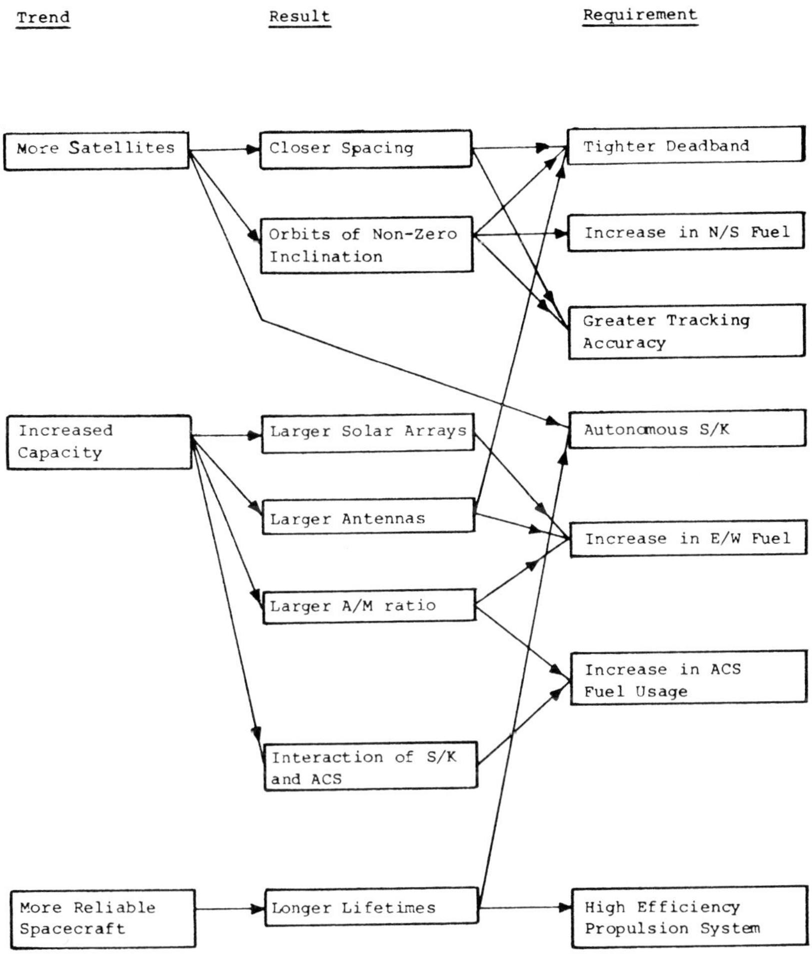

Fig.6.26 Future Trends

The feasibility of autonomous station-keeping has already been established, and an experimental system has been flown. The concept consists of determining the spacecraft's orbit on board the spacecraft using appropriate sensors and an ephemeris model stored in an on-board computer. Based upon a knowledge of the spacecraft's orbit and position, simple algorithms evaluate the time and magnitude of station-keeping manoeuvres. These manoeuvres are then initiated automatically. The system has been shown to be suitable for implementation in conjunction with electric propulsion systems.

Other possible future developments include a very large structure in geostationary orbit capable of being used as a *hanger* upon which to mount diverse communications payloads. This would reduce congestion on the orbit and the risk of collisions. The station-keeping of one large structure would replace that of many individual spacecraft, bringing about a reduction in ground station workload.

A summary pictorial representation of the impact of communications spacecraft evolution upon station-keeping requirements is shown in Figure 6.26.

6.6 Further Reading

A comprehensive review of station-keeping can be found in reference (a), which also contains a large list of references.

(a) Shrivastava, S.K. : 1978, "Orbital Perturbations and Station-Keeping of Communications Satellites". *J. of Spacecraft and Rockets*, Vol.15, No.2, March/April, pp.67–78.

East-West station-keeping and the associated perturbations are dealt with in references (b) and (c).

(b) Gideon, G.S. : 1969, "Tesseral Resonance Effects on Satellite Orbits", *Celestial Mechanics*, Vol.1, pp.167–189.

(c) Kamel, A., Eckman, D., and Tibbitts, R. : 1973, "East-West Station-Keeping Requirements of Nearly Synchronous Satellites Due to Earth's Triaxiality, and Luni-Solar Effects", *Celestial Mechanics*, Vol.8, pp.129–148.

North-South station-keeping and the associated perturbations are dealt with in references (d) and (e).

(d) Kamel, A. and Tibbetts, R. : 1973, "Some Useful Results on Initial Node Locations for Near-Equatorial Circular Satellite Orbits", *Celestial Mechanics*, Vol.8, pp.45–73.

(e) Gartrell, C.F. : 1983, "Time-Optimised North-South Station-Keeping", *J. Guidance*, Vol.6, No.3, pp.220–222.

In this chapter, the treatment of the physical mechanism of the perturbations counteracted by station-keeping, follows that developed by H.F. Michielsen and E.D. Webb, and presented in reference (f).

(f) "Station-Keeping of Geostationary Satellites Made Simple", Proceedings of the First Western Space Conference, Santa Monica, 1970, pp.704-727.

Some practical implementations of station-keeping are discussed in references (g), (h), and (i).

(g) Fagg, A.S. and MacLauchlan, J.B. : 1980, "Operational Experience on OTS 2", *J. of Guidance and Control*. Vol.4, No.5, pp.551-557.

(h) Pocha, J.J. and Edwards, A.F. : 1981, "L-Sat Station-Keeping", Proceedings of the Spacecraft Flight Dynamics Symposium, 18-22 May, ESA SP-160, pp.169-182.

(i) Soop, E.M. : 1985, "Geostationary Orbit Inclination Strategy", *ESA Journal*, Vol.9, pp.65-74.

The long term, uncontrolled evolution of the geosynchronous orbit is considered in reference (j).

(j) Agneni, A. et. al. : 1985, "A Useful Geometric Model for Luni-Solar Effects on the Long-Term Evolution of Geosynchronous Satellites", Paper AAS 85-321, presented at the AAS/AIAA Astrodynamics Specialist Conference, Vail, Colorado, Aug. 12-15, 1985.

Spacecraft cluster geometries and station-keeping are addressed in references (k) and (l).

(k) Walter, J.G. : 1982, "The Geometry of Satellite Clusters", *J. of the British Interplanetary Society*, Vol.35, pp.345-354.

(l) Murdoch, J. and Pocha, J.J. : 1982, "The Orbit Dynamics of Satellite Clusters", Proceedings of the 33rd International Astronautical Congress, Paris, IAF-82-54.

Autonomous station-keeping is covered extensively in the literature. A good reference is (m).

(m) Eckstein, M.C. and Leibold, A. : "Autonomous Station-Keeping of Geostationary Satellites", Spacecraft Pointing and Position Control, AGARD-AG-26.

CHAPTER 7

SPACECRAFT OPERATIONS

7.1 Introduction

In this chapter we shall discuss spacecraft operations, and how they are supported by the mission designer. Mission design precedes the operational phase, but its sole purpose is to ensure the success of the operational phase. Spacecraft operations are the active implementation of mission design. As the two activities are generally carried out by different people, close liaison and co-operation between them is essential.

The mission designer has to engage in several activities in support of spacecraft operations. For convenience, they can be divided into two groups: pre-launch support and post-launch support. During launch itself – this phase being defined as the period during which the spacecraft is attached to the launch vehicle – the mission designer is a spectator, and a helpless one at that, fervently hoping that the launch vehicle will do its job and that the satellite will be in the correct orbit.

Before we consider pre- and post-launch support in turn, a brief mention of the ground-to-space link will not be amiss. The spacecraft is controlled from the ground by monitoring its on-orbit behaviour and by commanding it to perform certain functions. This is achieved through the Telemetry, Tracking and Command (TTC) sub-system on board the satellite, and the associated hardware and software on the ground.

The spacecraft continually transmits to the ground a stream of information that tells the controllers the state the satellite is in. This includes data on the status and health of subsystems, the environment experienced by the spacecraft, and information allowing the attitude of the spacecraft to be determined. This data stream is known as the telemetry stream, and included within it is information allowing the position of the spacecraft to be determined on the ground. This process is known as tracking and is covered in detail in Chapter 9. The spacecraft responds to ground instructions by obeying commands sent by radio. The TTC subsystem on board the spacecraft receives these commands and routes them to the appropriate units on the spacecraft. The commands are then executed, and appropriate information is telemetered to the ground to enable the controllers to verify that the spacecraft has responded correctly.

7.2 Pre-Launch Activities

The mission designer provides pre-launch support in the following areas:

- interface activities with the launch vehicle authorities;
- definition of spacecraft-specific control software;
- assessment of the flight operations plan;
- participation in rehearsals for the operations activites;
- participation in negotiations with launch vehicle and control network authorities.

Let us now look at each of these activities.

7.2.1 Launch Vehicle Interface Activities

The mission designer is required to supply various items of data as inputs to documents forming agreements with the launch vehicle authorities. This data includes items such as the launch window, the characteristics of the orbit the vehicle is required to put the spacecraft into, the attitude of the spacecraft at separation from the launcher (if this is a selectable parameter), and the dynamical characteristics at separation (such as the spacecraft spin rate). The mission designer also receives information from the launch vehicle authorities that he must use during mission design.

7.2.2 Spacecraft-Specific Software

To control the spacecraft from the ground, the control centre requires an extensive suite of software. Some of this software is applicable to all the spacecraft controlled by the station, while other software is specific to either a single spacecraft or to a family of spacecraft. Within the former category would fall the following:

- orbit determination, orbit evolution and ephemeris generation;
- spacecraft control environment (mostly real-time software associated with the TTC link);

- data handling (file generation, archiving, graphics etc.);
- network control and co-ordination;
- hardware test and checkout.

This software would be an integral part of the spacecraft control facility, and would be made available for the support of any spacecraft controlled from that facility. Because spacecraft differ from each other in many details, however, this software needs to be augmented by spacecraft-specific software to form the complete suite necessary for control. Within the spacecraft-specific category would come the following software:

- spacecraft engineering data files;
- telemetry format composition;
- command composition;
- determination of manoeuvres required by the spacecraft;
- attitude determination;
- spin state determination;
- spacecraft sub-system management.

Different control centres vary in the division between general and spacecraft-specific software. The operations centre of the European Space Agency, provides as general software that which other centres would consider spacecraft-specific. Examples are: manoeuvre generation software and attitude determination software. Different control centres also vary in their approach towards the spacecraft-specific software. Some require full specifications of the software and use these to generate the software themselves. It is then integrated with the rest of the system. Others require the complete coding to be provided by the spacecraft authority, and will then integrate it into the rest of the system.

It can be seen that these variations involve large differences in the work that the spacecraft authority must undertake. The situation must, therefore, be carefully understood at an early stage, and the work planned and costed accordingly.

7.2.3 Assessment of the Flight Operations Plan

The flight operations plan regulates in detail the activities that constitute spacecraft control. It is prepared as a time-tagged schedule of events indicating the action to be taken at various times, the person responsible for undertaking the action, contingency procedures, etc. The plan is produced by the control centre personnel and is based upon operations requirements defined by the spacecraft designers. The mission designer is required to assess the flight operations plan in detail (checking for possible errors, adverse interactions. etc.), and finally give his approval.

7.2.4 Rehearsals and Training

The staff at the control centre who will be responsible for controlling the spacecraft require both training and rehearsals. They are required to understand, in broad terms, the spacecraft system characteristics, and the overall working of the spacecraft subsystems. This understanding is imparted through a short but intensive training programme prepared by the spacecraft authority, and within which the mission designer must play his part.

Training is followed by rehearsals of the various phases of the mission, wherein the activities detailed in the flight operations plan are gone through exactly as if a spacecraft were in orbit and being controlled. The rehearsals are planned by senior staff of the control authority and the spacecraft authority, and here again the mission designer is involved. The results of the rehearsals are assessed by a supervisory panel which then makes its recommendations. Sometimes this involves further rehearsals, and could even result in changes in the control procedures.

7.2.5 Negotiations and Liaison

During the life of a spacecraft project there is frequent and continuing liaison between the spacecraft user, the spacecraft designer, the launch vehicle authority, and the control and operations authority. The mission designer is required to support this activity, and ensure that the mission characteristics are communicated to those who require to know them. There is usually a well-defined plan of progress meetings and reviews, but *ad hoc* meetings are also arranged.

7.3 Post-Launch Activities

These are the activities that the previous training, preparations, and rehearsals are designed to facilitate. These activities result in the control of the spacecraft in orbit. The major objective of these activities is to establish the spacecraft correctly in its operational configuration, and to produce evidence that it is performing according to requirements.

Though the actual control of the spacecraft will be carried out by the control authority personnel, the spacecraft manufacturer is required to provide support at both the system and sub-system levels. The mission designer provides part of the system-level support. The purpose of this support is to monitor the mission, provide advice of a detailed nature when required, and, most important of all, to support the controllers when non-nominal behaviour occurs. This enables prompt and informed action to be initiated, and could have a crucial influence upon the continued well-being and useability of the spacecraft.

During the transfer orbit phase, when the control workload is heavy, intensive support is required. It is generally provided by two or three complete teams working shifts. After the spacecraft has performed its apogee manoeuvre and its operational attitude has been established, the control activities become less intensive, and the level of support can be correspondingly reduced. Once the spacecraft has been stationed at its operational longitude, the main control activity is station-keeping, sub-system and payload check-out, and commissioning.

Let us now outline the activities that occur during the various phases of the mission.

7.3.1 The Transfer Orbit Phase

During the transfer orbit phase the main activities are typically:

- the establishment of the TTC link;
- the determination of the state, and the monitoring of the health, of the spacecraft;
- the design, command, and verification of spin rate changes;
- tracking and orbit determination;
- the design of the apogee manoeuvre;
- the design, command, and verification of spacecraft attitude manoeuvres;
- the establishment of the correct operating mode of the various subsystems in response to environmental influences and possible equipment failures;
- the control and scheduling of the various stations within the ground network supporting the spacecraft;
- implementation and verification of the apogee manoeuvre.

7.3.2 The Drift Orbit Phase

After the apogee manoeuvre, the spacecraft will be in a near-synchronous orbit known as a drift orbit. Typically, the activities during this phase are:

- tracking and orbit determination;
- spacecraft health monitoring;
- the design, implementation, and verification of spin rate changes (if necessary);
- the implementation and verification of Sun acquisition;
- the deployment of the solar array panels (if appropriate);

- the implementation and verification of Earth acquisition;
- the configuration of the spacecraft in its *normal mode* of attitude stabilisation and control;
- the design and implementation of the station acquisition strategy;
- the design and implementation of station-keeping initialisation.

At the end of this phase, the spacecraft is in geostationary orbit at its operational position.

7.3.3 The Pre-Operational Phase

During this phase, the spacecraft's payload is activated and checked out, and the control parameters trimmed to maximise the spacecraft's operational lifetime. Typical activities are:

- tracking and orbit determination;
- station-keeping;
- payload switch-on, check-out and performance measurement;
- commissioning;
- the hand-over of the spacecraft to the user.

At the end of this phase, the spacecraft is ready for its operational phase.

7.3.4 The Operational Phase

The spacecraft's operational life begins with the hand-over of the spacecraft to the user. The spacecraft may then immediately be pressed into service, or it may be put into a *dormant* mode to serve as an on-orbit spare. Typical activities would be:

- tracking and orbit determination;
- spacecraft health monitoring;
- spacecraft life-time prediction;
- subsytem management (e.g. battery reconditioning);
- payload control and re-configuration;
- equipment re-configuration in response to failures;

- station-keeping;
- contingency manoeuvres.

The operations phase is the longest of all the phases and comprises the spacecraft's *useful* life.

7.3.5 The End-of-Life Phase

When the spacecraft lifetime prediction indicates that the station-keeping propellants are exhausted, or very soon will be, the spacecraft's operational life is at an end, and the end-of-life phase begins. This phase could include end-of-life equipment tests designed to yield information on how well the equipment has stood the test of time. Such tests provide valuable information for future design work. This phase could also include *graveyard* manoeuvres designed to remove the spacecraft from synchronous orbit and vacate its longitude slot. The slot can then be allocated to another spacecraft. This procedure minimises orbit over-crowding, and reduces the risk of collisions on the geostationary arc.

The activities could be:

- equipment tests and performance measurements;
- configuration of propulsion subsystem equipment for the *graveyard* burns;
- payload switch-off, retaining TTC capability;
- implementation and verification of the *graveyard* burns;
- venting of remaining fuel and pressurant;
- switch-off of remaining units (not TTC);
- switch-off of TTC.

The spacecraft is now uncontrollable and its operations are therefore at an end.

7.3.6 General Comments

It should be realised that spacecraft operations could change from one phase to another *and back again*. This could only realistically occur between the drift orbit phase, the pre-operational phase, and the operational phase. Consider the following scenario. Spacecraft *A*, operational at one longitude, is to be replaced by spacecraft *B*, while spacecraft *A* is required to operate at another longitude. In such a case, spacecraft *A* would revert from the operational phase to the drift orbit phase (during longitude change), then transition from the drift orbit phase to the pre-operational phase, and finally switch to the operational phase to undertake its new tasks. The control activities would be those associated with the current operating mode of the spacecraft.

Once the spacecraft is operational, very little support is required by the controllers. Support from the spacecraft manufacturer on an *as required* basis suffices. In the case of equipment failures, however, the level of support could rise dramatically, and would be dictated by the seriousness of the problem

CHAPTER 8

ORBIT PROPAGATION

8.1 Introduction

Orbit propagation consists of the prediction of the orbital characteristics of a satellite at some future date given the current orbital characteristics. Owing to a host of perturbations acting on a satellite, its orbital elements change with time. The perturbations that act upon a spacecraft can be divided into natural phenomena and those associated with satellite operations. Among the former are the effects of the Earth's gravitational potential, atmospheric effects, the gravitational attraction of the Sun, Moon, and Planets, etc., while the latter consist of the effects of spacecraft manoeuvres.

Orbit propagation, in some form or other, is carried out during most of the mission analysis activities. In many cases the problem can be simplified, and only the effects of the dominant perturbations considered. In this chapter we shall consider the subject in a more general way, and outline some of the methods used. The literature covering the perturbations that act upon an Earth-centred satellite, and the methods used to evolve the orbit, is extensive. Over the years a variety of systems of orbital elements has been developed, each system answering a particular specialised need, and having certain advantages over others in specific areas. In addition, various orbit integration methods have been developed in the search for greater accuracy, stability, and speed.

At this point, it is useful to define a perturbation. A perturbation is a deviation from some normal or expected motion. Contrary to the usual connotation of the word, it should not be supposed that perturbations are always small, for their effect can be as large as, or larger than, that of the primary attracting force. For example, many interplanetary missions would miss their targets entirely if the perturbing effects of other attracting bodies were not taken into account. Ignoring the effect of Earth oblateness on a low altitude satellite would cause a very large error in position prediction over a long period of time. Ignoring the triaxiality perturbation on a geostationary satellite would lead to a very large error in longitude prediction in due course.

There are two main categories of propagation techniques. These are referred to as *special perturbations* and *general perturbations*. *Special perturbation* techniques deal with the direct numerical integration of the equations of motion, including all the necessary perturbing accelerations. *General perturbation* techniques involve an analytic integration of series expansions of the perturbing accelerations. Here we shall be concerned with the former rather than the latter. Examples of *special perturbation* techniques are:

- Cowell's Method,
- Encke's Method,
- Variation of Parameters Method.

In the discussion that follows, we shall consider a number of perturbations of significance to communications spacecraft, two methods of evaluating their influence upon a spacecraft's orbit, and several methods of numerical integration that can be employed to propagate an orbit. In the treatment that follows, vector quantities are represented by bold type.

8.2 Orbit Perturbations

We shall now consider the various perturbations that act upon a satellite in orbit, emphasising those important for the orbits of geosynchronous spacecraft. The basic orbit will be considered to be a Keplerian conic, i.e. the orbit that results through the gravitational attraction of a spherical, perfectly homogeneous, central body. Perturbations will be considered to be those influences which cause a deviation from such an orbit. We shall consider the following perturbations:

- the Earth's gravitational potential,
- the gravitational influences of the Sun and Moon,
- aerodynamic drag,
- solar radiation pressure,
- spacecraft manoeuvres.

For an Earth-centred orbit the first three in the above list will act continuously and concurrently, though in geosynchronous orbit the aerodynamic drag is so small that it can be ignored, even in detailed calculations. The aerodynamic drag is important, however, during the parking and transfer orbits. Solar radiation pressure acts whenever the satellite is not in eclipse, which means that this perturbation is active for most of the satellite's lifetime. Perturbation during manoeuvres occurs only when the satellite's reaction control system is actuated, and is significant only when the manoeuvres are translational ones and not when they are purely rotational ones.

We shall now consider each in more detail.

8.2.1 The Earth's Gravitational Potential

The Keplerian conic occurs under the influence of the gravitational potential of a homogeneous, perfectly spherical, Earth. The Earth, however, is not a spherically symmetric body but is bulged at the Equator, flattened at the Poles, and generally asymmetric. Its gravitational potential is therefore a complex expression, which takes into account the *actual* mass distribution. For a homogeneous perfectly spherical Earth, the gravitational potential is given by:

$$u = \frac{\mu}{r}$$

For the Earth as it actually is, the expression for the gravitational potential can be written as:

$$u = \frac{\mu}{r}\left[1 - \sum_{n=2}^{\infty} J_n \left(\frac{R}{r}\right)^n P_n(\sin L) + \sum_{n=2}^{\infty}\sum_{m=1}^{n}\left(\frac{R}{r}\right)^n P_n^m(\sin L)\,(C_{nm}\cos m\lambda + S_{nm}\sin m\lambda)\right]$$

where: L = latitude,
λ = longitude,
r = distance of the satellite from the centre of the Earth,
R = Earth's mean radius.

Here $P_n(x)$ are the *Legendre functions* defined by:

$$P_n(x) = \frac{1}{2^n\, n!}\,\frac{d^n}{dx^n}\left[(x^2 - 1)^n\right] \qquad -1 \leqslant x \leqslant 1$$

and $P_n^m(x)$ are *associated Legendre functions* defined by:

$$P_n^m(x) = (1-x^2)^{\frac{m}{2}} \cdot \frac{d^m}{dx^m}\left[P_n(x)\right] \qquad -1 \leqslant x \leqslant 1\,;\, m \leqslant n$$

The constants J_n represent the *zonal* harmonics – that is, those harmonics which are dependent only on that mass distribution which is symmetric about the North–South axis of the Earth (they are not longitude dependent). The even-numbered harmonics are symmetric about the equatorial plane, and the odd-numbered harmonics are antisymmetric. The constants C_{nm} and S_{nm} represent *tesseral* harmonics – those dependent on both latitude and longitude.

Once the potential function u is known, the accelerations acting upon the satellite can be found. However we require the accelerations due to the perturbations only. Let the perturbation potential be u', then:

$$u = \frac{\mu}{r} + u'$$

Given u', the accelerations acting on a satellite can be found from:

$$\mathbf{f} = -\nabla u' = -\left[\mathbf{i}\,\frac{\delta u'}{\delta x} + \mathbf{j}\,\frac{\delta u'}{\delta y} + \mathbf{k}\,\frac{\delta u'}{\delta z}\right]$$

where **i**, **j**, **k** are unit vectors in the x, y, z directions respectively.

$$\therefore f_x = -\frac{\delta u'}{\delta x}\,;\quad f_y = \frac{\delta u'}{\delta y}\,;\quad f_z = -\frac{\delta u'}{\delta z}$$

These accelerations are in the form required by *Cowell's Method* and can be used directly. For the *Variation of Parameters Method*, the accelerations are required in the radial, f_1, transverse, f_2, and normal–to–orbit plane, f_3, directions. These are obtained through the following transformation:

$$f_r = -\frac{\delta u'}{\delta r} \quad \text{in the radially outwards direction;}$$

$$f_L = -\left(\frac{1}{r}\right)\frac{\delta u'}{\delta L} \quad \text{in the north direction at the subsatellite point;}$$

$$f_\lambda = -\left(\frac{1}{r\cos L}\right)\frac{\delta u'}{\delta \lambda} \quad \text{in the east direction at the subsatellite point.}$$

To convert these to orbital co–ordinate directions, (f_1, f_2, f_3), we require the azimuth angle (defined here as the angle North of the East direction) of the satellite velocity vector. If this angle is β, it is given by:

$$\tan\beta = \tan i \cos(\omega + f)$$

Now the acceleration components in orbital axes are given by:

$$f_1 = f_r$$

$$f_2 = f_\lambda \cos\beta + f_L \sin\beta$$

$$f_3 = f_L \cos\beta - f_\lambda \sin\beta$$

In this form, the perturbing accelerations can be incorporated directly into the *Variation of Parameters Method*. The expressions for f_r, f_L, and f_λ can be found in Appendix A.

8.2.2 Luni-Solar Perturbations

These are influences on the satellite of the gravitational attraction of the Sun and the Moon. The influences are evaluated separately, but follow the basic theory applicable to all third-body gravitational influences. The acceleration imposed upon the satellite by the gravitational attraction of a third body is given by:

$$\mathbf{f} = \mu_t \left(\frac{\mathbf{r}_{ts}}{r^3_{ts}} - \frac{\mathbf{r}_{te}}{r^3_{te}} \right)$$

where: μ_t = gravitational parameters of the third body,
$\mathbf{r}_{ts}$ = satellite-to-third body radiusvector (x_{ts}, y_{ts}, z_{ts}),
$\mathbf{r}_{te}$ = Earth-to-third body radius vector (x_{te}, y_{te}, z_{te}).

In Cartesian co-ordinates this becomes:

$$f_x = \mu_t \left(\frac{x_{ts}}{r^3_{ts}} - \frac{x_{te}}{r^3_{te}} \right);$$

$$f_y = \mu_t \left(\frac{y_{ts}}{r^3_{ts}} - \frac{y_{te}}{r^3_{te}} \right);$$

$$f_z = \mu_t \left(\frac{z_{ts}}{r^3_{ts}} - \frac{z_{te}}{r^3_{te}} \right)$$

In this form the expressions are suitable for use directly in Cowell's Method. In orbital axes – see Figure 8.1 – the perturbing accelerations are as follows:

$$f_1 = \frac{\mu_t}{\bar{R}^2_t} \left(\frac{(R_t/\bar{R}_t)\cos\rho_s \cos\sigma_s - (r/\bar{R}_t)}{(r_{ts}/\bar{R}_t)^3} - \frac{(R_t/\bar{R}_t)\ \cos\rho_s \cos\sigma_s}{(R_t/\bar{R}_t)^3} \right)$$

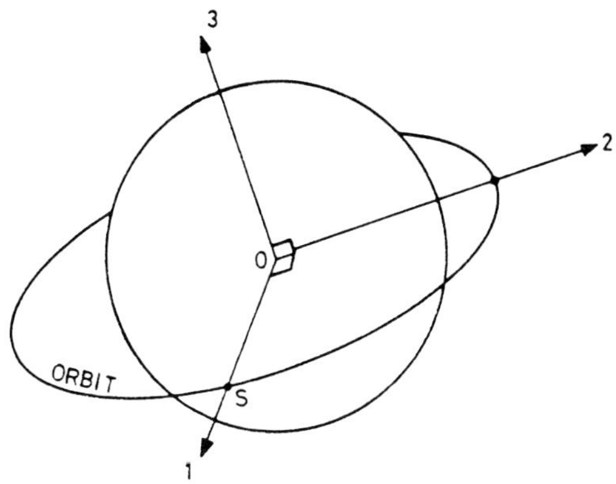

THE ORBITAL AXIS SYSTEM
AT THE POSITION OF THE SATELLITE, S
1 IS RADIAL (OUTWARDS)
2 IS TRANSVERSE (COMPLETES RIGHT-HANDED TRIAD)
3 IS NORMAL TO THE ORBIT PLANE (NORTHWARDS)

Fig.8.1 Definition of the Orbital Axes System

$$f_2 = \frac{\mu_t}{\bar{R}_t^2}\left(\frac{R_t}{\bar{R}_t}\right) \cos\rho_s \sin\sigma_s \left(\frac{1}{(r_{ts}/\bar{R}_t)^3} - \frac{1}{(R_t/\bar{R}_t)^3}\right)$$

$$f_3 = \frac{\mu_t}{\bar{R}_t^2}\left(\frac{R_t}{\bar{R}_t}\right) \sin\rho_s \left(\frac{1}{(r_{ts}/\bar{R}_t)^3} - \frac{1}{(R_t/\bar{R}_t)^3}\right)$$

where: R_t = instantaneous distance of the perturbing body from the centre of the Earth,
$\bar{R}_t$ = mean distance of the perturbing body from the centre of the Earth,
ρ_s and σ_s are defined in Figure 8.2,
r = distance of satellite from the centre of the Earth,
r_{ts} = third body-to-satellite distance.

The acceleration components f_1, f_2, and f_3 are in a form suitable for use in the *Variation of Parameters Method*.

The following constants apply:

$\bar{R}_t$ = 149597893 km for the Sun
$\bar{R}_t$ = 384400 km for the Moon

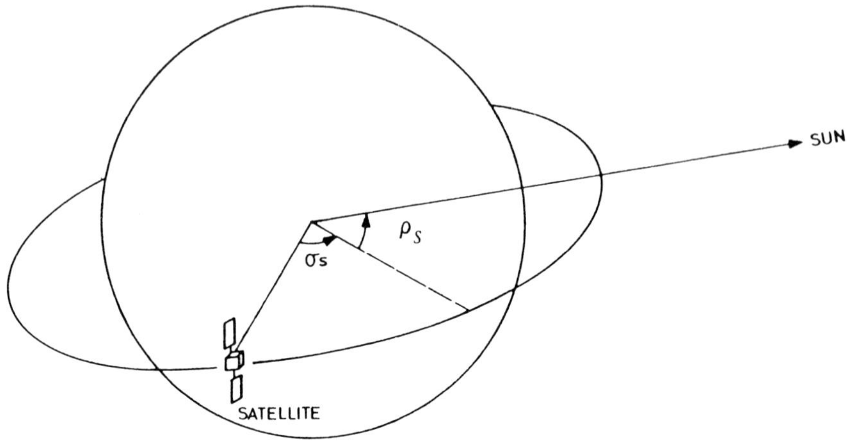

Fig.8.2 Definition of σ_S and ρ_S

$$\frac{\mu_t}{\bar{R}_t^2} = 15.930084387 \quad \text{km/sec}^2 \text{ for the Sun}$$

$$\frac{\mu_t}{\bar{R}_t^2} = 3.3179944 \times 10^{-8} \quad \text{km/sec}^2 \text{ for the Moon.}$$

8.2.3 Aerodynamic Drag

Owing to the non-zero density in the upper reaches of the atmosphere, the satellite suffers a drag force in a direction opposite to its motion. When the satellite is in geostationary orbit it can be assumed to be in a vacuum, and hence the drag force is zero. In parking and transfer orbits, however, drag effects cannot be neglected in an accurate orbit propagation calculation. The calculation of the drag force is, however, beset with uncertainties. This is due to several factors. The atmospheric density is imperfectly known and varies in a complex manner with time. There are daily, seasonal (several months), and long-period (several years), variations in addition to the variations with respect to altitude and latitude. An accurate atmospheric density model is in fact much too complex to be incorporated into most orbit evolution software and mean values are resorted to. The drag co-efficient is difficult to evaluate. It depends upon the satellite shape, the nature of its surfaces, and the area projected normal to the velocity vector. All three change with time, and once more mean values are generally used.

The expression for the acceleration due to drag is, however, a simple one:

$$\mathbf{f} = -\tfrac{1}{2}\rho \frac{A}{m} C_D V_a \dot{\mathbf{r}}_a$$

where:
- C_D = dimensionless drag coefficient associated with A,
- A = the cross-sectional area of the vehicle projected perpendicular to the velocity vector,
- m = satellite mass,
- ρ = atmospheric density encountered by the satellite,
- V_a = satellite velocity relative to the rotating atmosphere,
- $\dot{\mathbf{r}}_a$ = velocity vector of satellite relative to the rotating atmosphere.

The acceleration is then resolved into Cartesian co-ordinates for *Cowell's Method* as follows:

$$\dot{\mathbf{r}}_a = \begin{bmatrix} \dot{x} + \omega y \\ \dot{y} - \omega x \\ \dot{z} \end{bmatrix}$$

$\dot{\mathbf{r}} = (\dot{x}, \dot{y}, \dot{z})$ = satellite's inertial velocity
ω = rate of rotation of the Earth = 0.0000729211515 rad/sec

Therefore:

$$f_x = -\tfrac{1}{2}\rho \frac{A}{m} C_D V_a (\dot{x} + \omega y)$$

$$f_y = -\tfrac{1}{2}\rho \frac{A}{m} C_D V_a (\dot{y} - \omega x)$$

$$f_z = -\tfrac{1}{2}\rho \frac{A}{m} C_D V_a \dot{z} .$$

For the *Variation of Parameters Method* the acceleration components in the inertial frame, i.e. f_x, f_y, f_z, may be transformed into components in the orbital frame, i.e. f_1, f_2, f_3, through the inertial-to-orbital frame co-ordinate transformation.

8.2.4 Solar Radiation Pressure

The sunlit faces of a satellite receive radiation from the Sun. The impinging radiation causes a force on the satellite, proportional to the sunlit area, and in the opposite direction to the satellite-to-Sun direction. For communication satellites, with their large solar array area, this perturbation effect cannot be neglected when accurate orbit evolution is carried out. In the calculation, several simplifying assumptions are made:

- the reflectance of the satellite remains constant;
- radiation absorbed by the satellite is re-radiated equally in all directions, producing zero net force;
- satellite areas projected normal and parallel to the orbit plane remain constant (or mean values are used);
- solar parallax is ignored;
- lunar eclipses are ignored.

The satellite area projected perpendicular to the Sun–satellite line is considered in two parts. Area A is the area of the *sunlit* faces perpendicular to the orbit plane while Area B is the area of the *sunlit* faces parallel to the orbit plane. Note that for Earth-pointing satellites, an average value of A is required.

Let the Sun position be defined by (R_S, ρ_S, σ_S); where R_S is the Sun-Earth distance, ρ_S is the elevation angle of the Sun out of the orbit plane, and σ_S is the azimuth angle of the Sun measured from the satellite position, and in the orbit plane – as illustrated in Figure 8.2. Consider a face with cross-sectional Area A normal to the projection of the Sun direction in the orbit plane, i.e. the Sun zenith direction. The effective area, A, normal to the Sun line is:

$$A_1 = A \cos \rho_S$$

The incoming solar radiation, therefore, exerts a force per unit mass equal to $-PA_1 \cos \rho_S$ in the direction of the Sun zenith. The reflected radiation exerts a force per unit mass equal to $-PA_1 E \cos \rho_S$. Here P is the radiation pressure at the satellite, and E is the satellite's reflectivity. The resultant force per unit mass in the orbit plane, f_0, is therefore given by:

$$f_0 = -PA\,(1+E)\cos^2 \rho_S$$

Resolving this force into the radial direction (positive away from the Earth), and taking into account spacecraft mass, m, gives:

$$f_1 = \frac{P}{m}\, A\,(1+E)\cos^2 \rho_S \cos(-\sigma_S - \pi)$$

due to radiation incident on faces which are perpendicular to the orbit plane. Similarly by considering the area of the sunlit faces in the orbit plane, B, we derive the total perturbing acceleration in the radial direction.

$$f_1 = \frac{P}{m}\left[A\,(1+E)\cos \rho_S + B\,(1-E)\sin \rho_S\right] \cos \rho_S \cos(-\sigma_S - \pi)$$

and similarly for the transverse and normal directions we have:

$$f_2 = \frac{P}{m}\left[A\,(1+E)\cos\rho_S + B\,(1-E)\sin\rho_S\right]\quad \cos\rho_S\,\sin\,(-\sigma_S - \pi)$$

$$f_3 = -\frac{P}{m}\left[A\,(1-E)\cos\rho_S + B\,(1+E)\sin\rho_S\right]\quad \sin\rho_S$$

For the purpose of this simplified analysis, the reflectivity E is defined as:

$$E = \frac{\text{Reflected radiation}}{\text{Incident radiation}}$$

and P is taken to be $4.56 \times 10^{-6}\ \mathrm{N/m^2}$.

8.2.5 Spacecraft Manoeuvres

Except when a pure torque is applied, thruster firing will result in a perturbing acceleration on the orbit. Depending upon the nature of the manoeuvre, the thrust may be considered to be applied impulsively, or over a finite time interval.

If the thrust is not impulsive, it can be treated directly by resolving the thrust vector into components in the required directions. The perturbing accelerations, in the Cartesian frame, then become:

$$\ddot{x} = \frac{T_x}{m}\ ;\quad \ddot{y} = \frac{T_y}{m}\ ;\quad \ddot{z} = \frac{T_z}{m}$$

and can be used directly for the *Cowell Method*. For direct use in the *Variation of Parameters Method*, the accelerations due to the thrust must be derived in the orbital plane axes.

For impulsive thrusts the formulation is somewhat different. Impulsive thrusts are such that they occur over such a short interval of time that they can be considered to result in an instantaneous change in velocity. Below we shall derive the formulation for the *Variation of Parameters Method*.

Let the impulsive change in velocity be (V_1, V_2, V_3) in the orbital plane axes. Then, using the formulae derived in Section 8.4 we have:

$$g_i\ \Delta\theta = \frac{h^4}{\mu^3}\ V_i\ \frac{d\theta}{dt}\quad \text{for } i = 1 \text{ to } 3$$

Now $$\frac{dt}{d\theta} = h^3 k^2/\mu^2$$

$$\therefore \qquad g_i \, \Delta\theta = \frac{h^4}{\mu^3} \cdot \frac{\mu^2}{h^3 k^2} \, V_i = \frac{h}{\mu k^2} \, V_i$$

We know that:

$$\frac{dh}{d\theta} = hk^3 g_2$$

$$\therefore \qquad \Delta h = hk^3 g_2 \, \Delta\theta = \frac{h^2 k^3}{\mu k^2} \cdot V_2 = \frac{h^2 k}{\mu} V_2$$

For convenience let $B = hk/\mu$

$$\therefore \qquad \Delta h = hBV_2$$

And similarly for the other elements, we have:

$$\Delta\delta = \frac{h}{\mu} (V_1 \sin\theta + V_2 \cos\theta) + B(\delta + \cos\theta) V_2$$

$$\Delta\epsilon = \frac{h}{\mu} (V_2 \sin\theta - V_1 \cos\theta) + B(\epsilon + \sin\theta) V_2$$

$$\Delta l_0 = \tfrac{1}{2} B (l_1 \cos\theta + l_2 \sin\theta) V_3$$

$$\Delta l_1 = \tfrac{1}{2} B (l_0 \cos\theta - l_3 \sin\theta) V_3$$

$$\Delta l_2 = \tfrac{1}{2} B (l_0 \sin\theta + l_3 \cos\theta) V_3$$

$$\Delta l_3 = \tfrac{1}{2} B (l_1 \sin\theta - l_2 \cos\theta) V_3$$

The satellite mass after the impulsive manoeuvre must take into account the fuel used during the manoeuvre, i.e:

$$M_{after} = M_{before} \left(\exp \left(-\frac{\Delta V}{Ig} \right) \right)$$

where: $\Delta V = (V_1{}^2 + V_2{}^2 + V_3{}^2)^{1/2}$

g = acceleration due to gravity = 9.81 m/s^2
I = specific impulse of the propellant

The analysis gives direct increments to the transformation variables, and therefore no integration of the thrust need be performed.

Incorporation of impulsive thrusts in *Cowell's Method* is very much more simple. Since the numerical integration obtains, directly, the satellite's position and velocity at every point, the velocity increment due to an impulsive thrust is added directly to the satellite's velocity at the time at which the impulsive manoeuvre is executed, and the integration is started from the new initial conditions.

8.3 Cowell's Method

This is, conceptually, the simplest technique and involves the numerical integration of the satellite's equations of motion, with all the perturbations being considered as accelerations applied to the satellite. The main advantage of the method is the simplicity of formulation and implementation, and the fact that any number of perturbations can be handled simultaneously. But there are disadvantages also. When the motion is near a large attracting body, smaller integration steps must be used, which adversely affects execution time and accumulated error due to round-off.

Classically, and in most current applications, this method has been applied in a Cartesian co-ordinate system. However, it has been found that some improvement can be realised by formulating the problem in polar or spherical co-ordinates. In this case **r** will tend to vary slowly, and the angle change will often be monotonic. This will permit the use of larger step sizes for the same round-off error.

For the two-body problem with perturbations, the equation of motion can be written as:

$$\ddot{\mathbf{r}} + \frac{\mu}{r^3}\mathbf{r} = \mathbf{a}_p$$

where $\mathbf{r}$ = radius vector, $\mathbf{a}_p$ = perturbing acceleration, and μ = Earth's gravitational constant.

For numerical integration this would reduce to two first-order differential equations:

$$\dot{\mathbf{r}} = \mathbf{v}$$

$$\dot{\mathbf{v}} = \mathbf{a}_p - \frac{\mu}{r^3}\mathbf{r}$$

In cartesian co-ordinates this is further broken down into the following:

$$\dot{x} = V_x \; ; \; \dot{V}_x = a_{px} - \frac{\mu}{r^3} x$$

$$\dot{y} = V_y \; ; \; \dot{V}_y = a_{py} - \frac{\mu}{r^3} y$$

$$\dot{z} = V_z \quad ; \quad \dot{V}_z = a_{pz} - \frac{\mu}{r^3} z$$

where $r^2 = x^2 + y^2 + z^2$.

The perturbing acceleration, $\mathbf{a}_p$, is the vector sum of all the perturbations acting on the satellite. For example, if the effect of the Moon were to be considered, $\mathbf{a}_p$ would be given by:

$$\mathbf{a}_p = \mu_m \left(\frac{\mathbf{r}_{ms}}{r^3_{ms}} - \frac{\mathbf{r}_{me}}{r^3_{me}} \right)$$

where: μ_m = gravitational parameter of the Moon
$\mathbf{r}_{ms}$ = satellite-to-Moon radius vector
$\mathbf{r}_{me}$ = Earth-to-Moon radius vector.

Once the perturbing potential has been defined through an analytic formulation, the *state vector* (**r**, **v**) can be evolved by applying one of the many available numerical integration schemes. Once the state vector is obtained, the instantaneous orbit elements can be determined in a simple manner.

8.4 The Variation of Parameters Method

This was the earliest successful method of orbit propagation and was first developed by Euler in 1748. Whereas the *Cowell* and *Encke* methods are concerned with calculating the satellite's position and velocity directly, the *Variation of Parameters* method calculates the orbital elements or any other consistent set of parameters which adequately describes the orbit. A two-body orbit can be described by any consistent set of six parameters since it is described by three second-order differential equations. The classical orbital elements are only one of many possible sets. The essence of the method is to find how the selected set of parameters varies with time due to the perturbations. This is done by finding analytic expressions for the rate-of-change of the parameters in terms of the perturbations. These expressions are then integrated numerically to find their values at some later time. (Note that integrating the expressions analytically is the method of *general* perturbations). Since orbital elements vary very much more slowly than position and velocity, larger integration step-sizes are possible than in *Cowell's* or *Encke's* methods.

Here we shall describe the formulation using the classical orbital elements modified to remove singularities in inclination and eccentricity which arise when dealing with geosynchronous orbits.

The transformation equations are given by:

$$h^2 = a\,\mu\,(1 - e^2)$$

$$\delta = e \cos(\omega - \varphi)$$

$$\epsilon = e \sin (\omega - \varphi)$$

$$l_0 = \cos \frac{i}{2} \cos (\frac{\Omega + \varphi}{2})$$

$$l_1 = \sin \frac{i}{2} \cos (\frac{\Omega - \varphi}{2})$$

$$l_2 = \sin \frac{i}{2} \sin (\frac{\Omega - \varphi}{2})$$

$$l_3 = \cos \frac{i}{2} \sin (\frac{\Omega + \varphi}{2})$$

$$\varphi = f + \omega - \theta$$

where:
a = semi–major axis,
e = eccentricity,
i = inclination
Ω = right ascension of ascending node,
ω = argument of perigee,
f = true anomaly.

The independent variable is changed from time, t, to range angle, θ, measured from a reference point determined by φ. The reference point changes with time according to:

$$\frac{d\varphi}{dt} = - \cos i \, \frac{d\Omega}{dt}$$

To express the variation of the above parameters, i.e. h, δ, ϵ, l_0 etc., we define perturbing accelerations in the orbital plane co–ordinate system:

f_1 = radial component of acceleration,

f_2 = transverse component of acceleration,

f_3 = component of acceleration normal to the orbit plane.

These are illustrated in Figure 8.1 which also indicates the positive sense of these directions. We now express f_1, f_2 and f_3 as below, to simplify the writing of the equations:

$$g_i = \frac{h^4}{\mu^3} f_i \quad \text{for } i = 1 \text{ to } 3$$

$$k = (1 + \delta \cos\theta + \epsilon \sin\theta)^{-1}$$

The equations expressing the variation of the parameters in terms of the perturbing accelerations can now be written as:

$$\frac{dh}{d\theta} = h\,k^3 g_2$$

$$\frac{d\delta}{d\theta} = (g_1 \sin\theta + g_2 \cos\theta)\,k^2 + g_2\ (\delta + \cos\theta)\,k^3$$

$$\frac{d\epsilon}{d\theta} = (g_2 \sin\theta - g_1 \cos\theta)\,k^2 + g_2\ (\epsilon + \sin\theta)\,k^3$$

$$\frac{dl_0}{d\theta} = \tfrac{1}{2} g_3\,(l_1 \cos\theta + l_2 \sin\theta)\,k^3$$

$$\frac{dl_1}{d\theta} = \tfrac{1}{2} g_3\,(l_0 \cos\theta - l_3 \sin\theta)\,k^3$$

$$\frac{dl_2}{d\theta} = \tfrac{1}{2} g_3\,(l_0 \sin\theta + l_3 \cos\theta)\,k^3$$

$$\frac{dl_3}{d\theta} = \tfrac{1}{2} g_3\,(l_1 \sin\theta - l_2 \cos\theta)\,k^3$$

$$\frac{dt}{d\theta} = \frac{h^3 k^2}{\mu^2}$$

Additionally, we have an equation for the fuel flow rate of the propulsion system:

$$\frac{dm}{d\theta} = -\frac{F}{gI}\left(\frac{dt}{d\theta}\right)$$

For any perturbation being considered, analytic expressions are sought through which g_1, g_2, and g_3 can be computed. Then the equations are integrated with respect to the range angle θ. The parameters are then converted back to orbital elements using the expressions shown below:

$$a = h^2 / (\mu\,(1 - \delta^2 - \epsilon^2))$$

$$r = h^2 / (\mu\,(1 + \delta \cos\theta + \epsilon \sin\theta))$$

$$V^2 = \mu^2 \ (1 + 2\delta \cos\theta + 2\epsilon \sin\theta + \delta^2 + \epsilon^2)/h^2$$

$$e = \sqrt{\delta^2 + \epsilon^2}$$

$$f = \cot^{-1} \left[\frac{\delta \cos\theta + \epsilon \sin\theta}{\delta \sin\theta - \epsilon \cos\theta}\right]$$

$$\cos i = 1 - 2\,(l_1{}^2 + l_2{}^2)$$

$$\sin i = 2\,((l_1{}^2 + l_2{}^2)\,(l_0{}^2 + l_3{}^2))^{1/2}$$

$$\tan(\omega + f) = \frac{(l_1\,l_3 - l_0\,l_2)\cos\theta + (l_2\,l_3 + l_0\,l_1)\sin\theta}{(l_2\,l_3 + l_0\,l_1)\cos\theta - (l_1\,l_3 - l_0\,l_2)\sin\theta}$$

$$\tan\varphi = \frac{(l_1\,l_3 - l_0\,l_2)}{(l_0\,l_1 + l_2\,l_3)}$$

$$\tan\Omega = \frac{l_1\,l_3 + l_0\,l_2}{l_0\,l_1 - l_1\,l_3}$$

8.5 Orbit Integration Schemes

8.5.1 Introduction

In this section we shall consider the methods used to solve the orbit propagation equations. The technique of numerical integration is a complex one with a wide range of methods available to suit the problem being considered. Here we shall provide a brief description of some of the methods employed for orbit integration.

For orbit propagation to be carried out successfully, it is imperative that the integration method employed exhibits good characteristics in terms of accuracy and stability. The choice of the method must depend on the formulation of the equations, and upon answers to questions such as:

- Will a large number of integration steps be required?
- Is the formulation sensitive to round–off errors?
- Will a constant step–size be adequate?
- Does the error of the estimate need to be quantified?

For the integration scheme to be efficient, the factors that must be considered are, speed, stability, accuracy, and storage requirements.

Basically, the orbit propagation problem is that of solving numerically the initial value problem:

$$\frac{dy_1}{dt} = f_1 \ (t, y_1, y_2, y_3 \ldots\ldots y_n)$$

$$\frac{dy_2}{dt} = f_2 \ (t, y_1, y_2, y_3 \ldots\ldots y_n)$$

$$\frac{dy_n}{dt} = f_n \ (t, y_1, y_2, y_3 \ldots\ldots y_n)$$

for $a \leqslant t \leqslant b$

subject to the initial conditions:

$$y_1\,(a) = \beta_1,\ y_2\,(a) = \beta_2\ ,\ \ldots\ldots y_n\,(a) = \beta_n$$

Numerical solution of this sytem will result in approximations obtained at a number of discrete *mesh points* within the interval (a, b). Once the approximate solution has been obtained at the mesh points, the approximate solution at other points is obtained by interpolation procedures.

The various methods that have been evolved for the solution of these equations all utilise Taylor's polynomial expansion and the associated *Taylor Theorem*. *Taylor's Theorem* is as follows.

If f is continuous in (a, b) and $f^{(n+1)}$ exists in (a,b), then for x_0 in (a,b), for every x in (a, b) there exists a value ϕ between x_0 and x such that:

$$f(x) = P_n\,(x) + R_n\,(x)$$

with $$P_n(x) = f(x_0) + f'\,(x_0)\,(x-x_0) + \frac{f''\,(x_0)}{2!}(x-x_0)^2 + \ldots\ldots \frac{f^n(x_0)}{n!}\,(x-x_0)^n$$

$$= \sum_{k=0}^{n} \frac{f^{(k)}(x_0)}{k!}\,(x-x_0)^k$$

and $$R_n(x) = \frac{f^{(n+1)}(\phi)}{(n+1)!}\,(x-x_0)^{n+1}$$

Here $P_n(x)$ is called the nth-degree Taylor polynomial for f about x_0 and $R_n(x)$ is called the remainder term (or truncation error) associated with $P_n(x)$. The truncation error is the error incurred by considering only a finite number of terms in the infinite Taylor series expansion. The greater the number of terms in the Taylor polynomial, the smaller the truncation error.

8.5.2 Runge-Kutta Methods

Taylor methods, used directly, have the disadvantage that high-order derivatives of f must be evaluated. This can be a complicated and time-consuming process, and, therefore, Taylor methods are seldom used in practice. *Runge-Kutta methods* have the high-order local truncation error of the Taylor methods, but avoid the consideration of the derivatives of $f(t,y)$. We shall state the *Fourth-Order Runge-Kutta method* without proof. To remind outselves that the solution is only an approximation of the actual value, we shall use the notation:

$$\frac{dy}{dt} = f(t,y)$$

$$w_0 = y(a) = \beta \; ; \; w_i = y(t_i) \; ; \; w_{i+1} = y(t_{i+1})$$

We then have:

$$w_{i+1} = w_i + \frac{1}{6}(k_1 + 2k_2 + 2k_3 + k_4)$$

where:

$$k_1 = h\, f(t_i, w_i)$$

$$k_2 = h\, f(t_i + \frac{h}{2}, \; w_i + \frac{k_1}{2})$$

$$k_3 = h\, f(t_i + \frac{h}{2}, \; w_i + \frac{k_2}{2})$$

$$k_4 = h\, f(t_{i+1}, \; w_i + k_3)$$

$$h = \text{step-size} = t_{i+1} - t_i$$

This method has local truncation error $0(h^4)$ provided the solution $y(t)$ has five continuous derivatives. The method is a single-step method in that only information on the previous step is used for the estimation of the next step. Furthermore, the step size is fixed. The main computational effort is in the evaluation of f. The fourth-order method requires four evaluations per step. An advantage is that no start-up procedure is necessary.

A variation of this method, known as the *Runge-Kutta-Gill method*, has been developed for high-speed digital computers to use a minimum of storage registers, and to control the growth of round-off errors. The formulae are:

$$w_{i+1} = w_i + \frac{1}{6}k_1 + \frac{1}{3}(1 - \sqrt{\tfrac{1}{2}})k_2 + \frac{1}{3}(1 + \sqrt{\tfrac{1}{2}})k_3 + \frac{1}{6}k_4$$

where: $k_1 = h\,f(t_i, w_i)$

$$k_2 = h\,f(t_i + \frac{h}{2},\; w_i + \frac{k_1}{2})$$

$$k_3 = h\,f(t_i + \frac{h}{2},\; w_i + (\sqrt{1/2} - 1/2)\,k_1 + (1 - \sqrt{1/2})\,k_2)$$

$$k_4 = h\,f(t_{i+1}, w_i - \sqrt{1/2}\,k_2 + (1 + \sqrt{1/2})\,k_3)$$

$$h = t_{i+1} - t_i$$

8.5.3 Runge-Kutta-Fehlberg Method

A disadvantage with the above methods is that they do not permit error control within a specified limit. It would be a significant advantage to have a method:

$$w_{i+1} = w_i + h_i\,\phi\,(t_i, h_i\,w_i)$$

for approximating the solution $y(t)$ to the initial value problem

$$\frac{dy}{dt} = f(t, y) \quad , \quad a \leqslant t \leqslant b \quad , \quad y(a) = \beta$$

which had the property that the minimum number of mesh points were used to ensure that the global error $|\,y\,(t_i) - w_i\,|$ did not exceed a specified tolerance, ϵ, for any of the mesh points. The *Runge-Kutta-Fehlberg method* provides just such a facility. Here the step size is monitored, and appropriately adjusted to ensure that the estimation error is always within the specified tolerance. The method consists of using a *Runge-Kutta method* of order five,

$$w_{i+1} = w_i + \frac{16}{135}\,k_1 + \frac{6656}{12825}\,k_3 + \frac{28561}{56430}\,k_4 - \frac{9}{50}k_5 + \frac{2}{55}k_6$$

to estimate the local error in a *Runge-Kutta method* of order four;

$$w_{i+1} = w_i + \frac{25}{216}\,k_1 + \frac{1408}{2565}\,k_3 + \frac{2197}{4104}\,k_4 + \frac{1}{5}\,k_5$$

where: $k_1 = h\,f(t_i, w_i)$

$$k_2 = h\,f(t_i + \frac{h}{4},\; w_i + \frac{k_1}{4})$$

$$k_3 = h\,f\left(t_i + \frac{3}{8}h,\ w_i + \frac{3}{32}k_1 + \frac{9}{32}k_2\right)$$

$$k_4 = h\,f\left(t_i + \frac{12}{13}h,\ w_i + \frac{1932}{2197}k_1 + \frac{7200}{2197}k_2 + \frac{7296}{2197}k_3\right)$$

$$k_5 = h\,f\left(t_i + h,\ w_i + \frac{439}{216}k_1 - 8k_2 + \frac{3680}{513}k_3 - \frac{854}{4104}k_4\right)$$

$$k_6 = h\,f\left(t_i + \frac{h}{2},\ w_i - \frac{8}{27}k_1 + 2k_2 - \frac{3544}{2565}k_3 + \frac{1859}{4104}k_4 - \frac{11}{40}k_5\right)$$

The error control procedure uses h as the step size in the derivation of $\overline{w}_{i+1}$ and w_{i+1}. These two values are then used to determine what the step size should be to satisfy the error tolerance thus:

$$q = \left(\frac{\epsilon h}{2\ |\overline{w}_{i+1} - w_{i+1}|}\right)^{1/4}$$

and the new step size is qh.

This value of step size is then used to repeat the current step if the error tolerance is violated, or as an estimate of the step size for the next step in the integration, if it is not. In practice the value of q is conservatively chosen to minimise the re-calculations of the function for the revised step size.

8.5.4 Multi-Step Methods

The methods discussed previously are one-step methods in that only information from the one previous step is used in the estimate for the next. The approximate solution is, however, available at each of the previous mesh points, and these can be expected to be *more* accurate, since the error increases as the integration steps onwards in time. It would be desireable, therefore, to use this information for the estimation of the next value. Such methods have been developed, and are called multi-step methods.

One such method is the *Adams-Bashforth fourth-order method*. The following equations apply:

$$w_{i+1} = w_i + \frac{h}{24}\left[55\,f(t_i, w_i) - 59f(t_{i-1}, w_{i-1}) + 37f(t_{i-2}, w_{i-2})\right.$$
$$\left. - 9f(t_{i-3}, w_{i-3})\right]$$

with $w_0 = \beta_0, w_1 = \beta_1, w_2 = \beta_2$ and $w_3 = \beta_3$

This method is known as an *explicit* method since the right-hand side of the difference equation is independent of w_{i+1}. The local truncation error of the method is given by:

$$\tau_{i+1} = \frac{251}{720} y^{(5)} (\mu_i) h^4$$

where $y^{(5)} = \dfrac{d^5 y}{dt^5}$ and $t_i < \mu_i < t_{i+1}$.

Another method is the *Adams-Moulton fourth-order method* expressed by the following:

$$w_{i+1} = w_i + \frac{h}{24} [9 f(t_{i+1}, w_{i+1}) + 19 f(t_i, w_i) - 5 f(t_{i-1}, w_{i-1}) + f(t_{i-2}, w_{i-2})]$$

with $w_0 = \beta_0, w_1 = \beta_1$, and $w_2 = \beta_2$.

This method is known as an *implicit* method because w_{i+1} appears on both sides of the difference equation, and, therefore, can only be obtained implicitly. This is a significant disadvantage, as sometimes it is very difficult or even impossible to make w_{i+1} the subject of the formula. The local truncation error is given by:

$$\tau_{i+1} = -\frac{19}{720} y^{(5)} (\mu_i) h^4 \, .$$

It can be seen that this method is more accurate than the *Adams-Bashforth method*.

It should be noted that both the above methods require a solution over the first three or four mesh points before they can be implemented, i.e. a starting procedure is required. This may be supplied by a *Runge-Kutta method* of comparable accuracy.

8.5.5 Predictor-Corrector Methods

In practice, multi-step methods are better used in the predictor-corrector mode. Here an *explicit* method provides the prediction while the *implicit* method provides the correction to the estimate. A typical implementation could be as follows.

For a fourth-order method, the *Runge-Kutta* technique is used to calculate the starting values w_0, w_1, w_2 and w_3. Then the *Adams-Bashforth method* is used to calculate $w_4^{(p)}$ from:

$$w_4^{(p)} = w_3 + \frac{h}{24}[55f(t_3, w_3) - 59f(t_2, w_2) + 37f(t_1, w_1) - 9f(t_0, w_0)]$$

This approximation to $y(t_4)$ is then used in the *Adams-Moulton method* to calculate $w_4^{(c)}$ from:

$$w_4^{(c)} = w_3 + \frac{h}{24}[9f(t_4, w_4^{(p)}) + 19f(t_3, w_3) - 5f(t_2, w_2) + f(t_1, w_1)]$$

The value of $w_4^{(c)}$ is then the approximation to $y(t_4)$. The technique of using the *Adams-Bashforth method* to evaluate the predictor, and the *Adams-Moulton method* to compute the correct estimate, is repeated for each successive step of the iteration. Each step requires five function evaluations as opposed to three for the *Adams-Moulton method*. The problem of obtaining an explicit relation for w_{i+1} has, however, been avoided altogether. *Predictor–corrector methods*, when used in conjunction with step size regulation are powerful and accurate tools for the numerical integration of initial value problems. Below we shall outline how step-regulation can be accomplished.

For a given tolerance of ϵ on the allowed error of the estimate, for any step $i+1$ we evaluate:

$$\sigma = \frac{|w_{i+1}^{(c)} - w_{i+1}^{(p)}|}{10h}$$

If σ satisfies the relation:

$$\frac{\epsilon}{10} < \sigma < \epsilon$$

the value $w_{i+1}^{(c)}$ is accepted, and the integration moves forward one step using the step size h. If σ is not in this range, the step size is altered to qh, where $q > 0$, and is obtained from the relation

$$q = \left(\frac{\epsilon}{2\sigma}\right)^{1/4}$$

The value for w_{i+1} is now calculated for the new step size, qh, and the process repeated until σ satisfies the inequality relationship above. Then the integration moves forward one step with step size qh. Note that since multi-step methods require equal step sizes for the starting values, any change in step size necessitates recalculating new starting values at that point. The algorithms are therefore designed to minimise the number of step size changes.

An important attribute of a numerical integration technique is stability. A method is considered stable if small changes in the initial conditions produce correspondingly small changes in the subsequent approximations. Since for orbit propagation, the initial values are themselves estimates, stability is a very important requirement in the integration technique used to propagate the orbit. It can be shown that both the Adams–Bashforth and Adams–Moulton methods are strongly stable, and hence so is a predictor–corrector scheme based on them.

A useful comparison of various integration methods is shown in Table 8.1, taken from NASA SP33 – "Orbital Flight Handbook".

8.5.6 Round-Off Error

In the discussion above we have considered various numerical integration methods with a view to their efficiency, convenience, stability, and truncation error. There is a further source of error, called round–off error, which arises out of the way numbers are manipulated in a digital computer. This error is unavoidable and can only be minimised by careful design of the algorithm and the computer code.

A digital computer cannot represent any real number exactly, but only to a certain accuracy. This is because a number is represented by only a finite number of bits, giving accuracy to only a finite number of digits. There is also a limit to the size of number a computer can represent. One way to reduce the effect of round–off is to increase the number of bits allocated to the representation of the number. Most computers facilitate this through double–precision and even multiple–precision arithmetic. In the former, double the number of bits are allocated to the representation of a number. The penalty with this approach is that computation time and storage increase rapidly and hence so does the cost of computing. Other, more subtle methods are also available. These can be summarised by the following guidelines:

- avoid subtracting nearly equal numbers,
- avoid dividing by very small numbers,
- avoid multiplying by very large numbers,
- order the computation to reduce the number of computations, specially multiplications.

With careful coding very substantial improvements in accuracy and stability are possible.

Table 8.1 Comparison of Integration Methods (From NASA SP-33, Vol.1, Part 7)

Method of Numerical Integration	Truncation Error	Ease of Changing Step-Size	Speed	Stability	Round-off Error Accumulation
Single Step Methods					
Runge-Kutta	h^5	*	Slow	Stable	Satisfactory
Runge-Kutta-Gill	h^5	*	Slow	Stable	Satisfactory
Bowie	h^3	Trivial (step-size varied by error control)	Fast	Stable	Satisfactory
Fourth Order Multi-Step Predictor-Corrector					
Milne	h^5	Excellent	Very fast	Unstable	Poor
Adams-Moulton	h^5	Excellent	Very fast	Unconditionally stable	Satisfactory
Higher Order Multi-Step					
Adams Backward Difference	Arbitrary	Good	Very fast	Moderately stable	Satisfactory
Gauss-Jackson**	Arbitary	Awkward and expensive	Fast	Stable	Excellent
Obrechkoff	h^7	Excellent	***	Stable	Satisfactory
Special Second Order Equations [$z = f(t,z)$]					
Special Runge-Kutta	h^5	*	Slow	Stable	Satisfactory
Milne-Stormer	h^6	Excellent	Very fast	Moderately stable	Poor

* R-K (single-step) trivial to change steps, very difficult to determine proper size.
** Gauss-Jackson is for second order equations.
*** Speed of Obrechkoff depends on complexity of the higher order derivatives required; it could be very fast.

8.6 Further Reading

A brief treatment of orbital perturbations is found in:

(a) Bate, R.R., Mueller, D.D., and White, J.E. : "Fundamentals of Astrodynamics". Published by Dover Publications, 1971.

More detailed treatments are found in:

(b) Roy, A.E. : "Foundations of Astrodynamics". Published by Macmillan Company, New York, 1965.

(c) Herrick, S. : "Astrodynamics", Vols.1 and 2. Published by Van Nostrand, Reinhold Co., London, 1972.

Perturbations due to aerodynamic drag are discussed in the following references:

(d) de Lafontaine, J. and Garg, S.C. : 1980, "A Review of Satellite Lifetime and Orbit Decay Prediction", UTIAS Review No.43.

(e) Fredo, R.M. and Kaplan, M.H. : 1981, "Procedure for Obtaining Aerodynamic Properties of Spacecraft", *J. of Spacecraft and Rockets*, Vol.18, No.4, pp.367–373.

(f) Knuth, E.L. : 1980, "Free-Molecule Normal-Momentum Transfer at Satellite Surfaces", *AIAA Journal*, Vol.18, No.5, pp.602–605.

(g) Santora, F.A. : 1976, "Drag Perturbations of Near-Circular Orbits in an Oblate Diurnal Atmosphere", *AIAA Journal*, Vol.14, No.9, pp.1196–1200.

(h) Perini, L.L. : 1975, "Economical Scheme for Estimating Orbital Lifetimes", *J. of Spacecraft and Rockets*, Vol.12, No.6, pp.323–324.

(i) King-Hele, D.G. : 1964, "Theory of Satellite Orbits in an Atmosphere". Butterworth Publishing Co., London.

(j) Sterne, T.E. : 1959, "Effect of the Rotation of a Planetary Atmosphere Upon the Orbit of a Close Satellite", *ARS Journal*, October, pp.777–782.

Derivation of the harmonics of the Earth's Potential Field is found in the following:

(k) Kaula, W.M. : 1967, "Review of Tesseral Harmonics Determination from Satellite Orbit Dynamics". Proceedings of the II International Symposium on the Use of Artificial Satellites for Geodesy, edited by G. Veis; Athens, Greece, pp.133–147.

(l) Kaula, W.M. : 1971, "Geopotential and Geodetic Models", *Astrodynamics and Celestial Mechanics*, AIAA Selected Reprint Series, Vol.XII, edited by V.G. Szebehely, pp.40–60.

(m) Gaposchkin, E.M. : 1974, "1972 Smithsonian Institute Standard Earth", *Journal of Geophysical Research*, Vol.79, pp.5377–5411.

An excellent account of integration methods is found in:

(n) Burden, R.L., et. al. : 1981, "Numerical Analysis", Second Edition, Published by Prindle, Weber and Schmidt, Boston, Mass.

CHAPTER 9

TRACKING AND ORBIT DETERMINATION

9.1 Introduction

The in–orbit control of a spacecraft requires knowledge of its position in space. For geostationary spacecraft the position is required relative to Earth–centred axis systems, both inertially fixed and rotating with the Earth, the former largely for manoeuvre design and event prediction, the latter for purposes of Earth station coverage and antenna pointing.

The process by which the spacecraft's position is determined is known as *tracking and orbit determination*. *Tracking* is the process of making observations of the spacecraft's position relative to a tracking station or other fixed point (such as a star) whose position is accurately known. *Orbit determination* is the process wherein the tracking observations are used to determine the spacecraft's orbital characteristics and its position in space.

This chapter will discuss the techniques used for the tracking and orbit determination of geosynchronous spacecraft, and the facilities generally available to perform these functions. Before we do this, however, some comments upon the mission characteristics of geostationary spacecraft with respect to tracking will be useful to the reader.

As far as tracking is concerned, the spacecraft's mission can be divided into the transfer orbit phase and the geosynchronous orbit phase. The tracking characteristics for the two phases are generally quite different.

In the transfer orbit phase the spacecraft is only intermittently visible from any one ground station. Moreover, the command and control workload is rather heavy, and is compressed into a relatively short time span. These considerations require the spacecraft's orbit to be determined to acceptable accuracy (see Section 3.4) quite rapidly. This in turn implies that several ground stations are required to co–operate in the process of tracking and orbit determination. Such a collection of co–operating ground stations is known as a *network*, and is characterised by commonality of tracking systems and standards. Overall control is exercised by a single element of the network, known as the *control centre*. The ground stations within the network respond to tracking measurement requests from the control centre, and pass on tracking measurements to the control centre. At the control centre, the tracking measurements from all the stations are collected, and the orbit determination process is carried out. The tracking system (mainly the choice of frequencies) used in transfer orbit is intimately connected with the telemetry and command system that is employed during this phase. In fact, the telemetry, tracking, and command (TTC) hardware on board the spacecraft is selected for compatibility with the ground station network chosen to support the spacecraft during its transfer orbit phase.

When the spacecraft is operational in geostationary orbit, tracking is generally carried out through the communications payload equipment, and must therefore conform with its frequency band. Since the spacecraft is continuously within view of the ground station used to communicate with it, a network is not required. Futhermore, since the schedule of activities is very much more relaxed than in transfer orbit, tracking is required infrequently and is generally a more long drawn out process, in that observations are collected over a day or two before the orbit is determined. The transfer orbit TTC system acts as the back-up system during the operational phase of the spacecraft.

Let us now consider tracking techniques and orbit determination methods.

9.2 Tracking

9.2.1 Types of Measurement

9.2.1.1 Range

The most common type of tracking measurement for geostationary spacecraft is ranging. Ranging is the determination of the spacecraft-to-ground station distance. This is done by bouncing a coded signal off the spacecraft and detecting it at the ground station. The signal could consist of either a transmission at radio frequency, or a laser beam. Only the former is in use for geostationary spacecraft.

The ranging process can be broken down into the discrete steps outlined below.

(1) transmission of the signal by the ground station;

(2) propagation of the signal from the ground station to the spacecraft;

(3) signal turn-around by the spacecraft;

(4) propagation of the signal from the spacecraft to the ground station;

(5) reception of the signal by the ground station.

Errors arise at each step of the process, instrumentation errors for steps (1), (3), and (5), and environmental effects for steps (2) and (4).

Environmental Effects

Let us first consider the environmental effects which influence a radio frequency transmission as it traverses the Earth's atmosphere. The Earth's atmosphere can be considered to be divided into layers as shown in Figure 9.1. The *troposphere* extends from the Earth's surface up to an altitude of about 35 km. Above this there is the *free-space* region extending up to about 85 km. Then we have the *ionosphere* which extends to an altitude of about 1000 km.

Under normal propagation conditions, the refractive index of the atmosphere decreases with height so that a signal travels more slowly near the ground than at higher altitudes. This variation in velocity results in the bending of the signal path as shown in Figure 9.1.

The tropospheric range error is due principally to the fact that the ray *propagation velocity* in the troposphere is different from the speed of light in vacuum. A second order effect is the physical path length difference between the ray path and the line-of-sight path. The ray *propagation velocity* is dependent upon the refractivity of the medium and in the troposphere is less than the speed of light in vacuum. This leads to an increase in the time delay between the transmitted and received signals, which in turn translates to an overestimate of the true range. From realistic refractivity models of the atmosphere, a correction can be made for this effect, and it has been shown that up to 97% of the range error can thus be accounted for. For elevation angles greater than about 5°, the residual range error, i.e. the error remaining after the correction has been applied, is negligible.

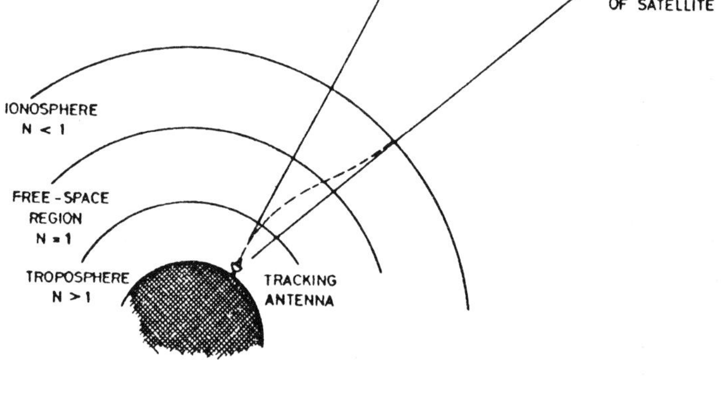

Fig.9.1 Range Errors Due to the Earth's Atmosphere

The measured elevation angle of the satellite is also in error due to the refraction within the troposphere, the effect being more pronounced at small elevation angles. Once more, models of the troposphere permit corrections to be made, and the residual elevation angle error, even for fairly crude corrections, can be reduced to the order of 0.5 mrad for elevation angles greater than 5°.

In the range of 3 MHz to 30 GHz, it has been shown that tropospheric refraction effects are independent of the carrier frequency, i.e. the frequency of ray propagation.

In the ionosphere, the refractive index, N, is defined as the ratio of the velocity of light in vacuum to the *phase velocity*. This definition is similar to the one applicable to the troposphere except that the *phase velocity* replaces the *propagation velocity*. As the *phase velocity* is dependent upon frequency, the ionosphere is known as a dispersive medium. In the ionosphere, the *phase velocity*, in absolute terms, is greater than the speed of light in vacuum. The signal velocity is associated with the velocity of propagation of the envelope of a modulated sinusoid. Since a pulse consists of a spectrum of frequencies, phase distortion is inherent in dispersive media. If the pulse spectrum is narrow compared to the carrier frequency, however, an average signal or *group velocity* can be calculated, which can be used to assess the time delays of propagation through the ionosphere. It can be shown that the *group velocity* is less than the speed of light in vacuum, and hence a time delay is introduced. The effect is, therefore, additive to the tropospheric affect, and leads to an overestimate of the range of the spacecraft from the ground station.

In principle, the ray-bending or refraction mechanism within the ionosphere is similar to that in the troposphere, but because the refractive index is less than unity, the ray bends in the opposite sense to that in the troposphere. This leads to some mitigation of the elevation angle error due to refraction within the atmosphere. The ray-tracing method of estimating elevation angle error is still applicable to the ionosphere, but the refractivity model is currently so inadequate, that reliable corrections are not possible. The correction is dependent upon the stability of that portion of the ionosphere through which the measurement ray passes, and since the ionospheric layers are always in a state of motion – especially around dawn and dusk – accurate corrections are difficult to arrive at.

It has been found that the lower the carrier frequency, the greater the elevation angle and range errors. Furthermore, the errors increase with the height of the satellite.

For ECS and MARECS, which use VHF ranging during the transfer orbit phase, tracking errors due to ionospheric effects dominate.

Instrumentation Errors

Let us now consider errors due to imperfect instrumentation. The instrumentation, like the atmosphere, gives rise to errors in the range by causing a delay when the signal passes through any component of the system. These delays are inherent in any piece of equipment, and tend to remain constant or slow-changing over a set of tracking measurements. Consequently these are treated as bias errors in the tracking and orbit determination process. In addition, an error arises from inaccuracies in the clock used for measuring transmission and reception times. This effect, however, is negligible in comparison with the other errors, and is subject to regular calibration.

Finally, another error present in the tracking system is the uncertainty in the ground station location with respect to the geoid, and is due mainly to errors in surveying the location of the ground station antenna. Typical errors are 20 m about each of the three axes. Decimetre accuracies are possible if the station location has been determined by laser-ranging of special low-Earth satellites.

Returning to the single ground station ranging case, the two-way range R_2, measured by tracking, can be expressed as follows:

$$R_2 = \Delta R_T + R_{GS} + \Delta R_{GS} + \Delta R_S + R_{SG} + \Delta R_{SG} + \Delta R_R$$

where: R_{GS} is the actual up-link range,
R_{SG} is the actual down-link range,

and systematic errors arise from the following sources:

ΔR_T = instrumentation delays on transmission from ground station,

ΔR_{GS} = atmospheric effects on the up-link,

ΔR_S = instrumentation delays at the spacecraft,

ΔR_{SG} = atmospheric effects on the down-link,

ΔR_R = instrumentation delays on reception by the ground station.

Assuming that over the period of one measurement,

$$\Delta R_T = \Delta R_R$$

$$\Delta R_{GS} = \Delta R_{SG}$$

$$R_{GS} = R_{SG},$$

the measured range ρ_M, of the spacecraft from the station is given by:

$$\rho_M = R_{GS} + \Delta R_{GS} + \Delta R_T + \tfrac{1}{2}\, \Delta R_S$$

and therefore the total systematic error on the measurement is:

$$\rho_M - R_{GS} = \Delta R_{GS} + \Delta R_T + \tfrac{1}{2}\, \Delta R_S$$

In addition to systematic errors, there are also *noise* errors. As the name suggests, these occur randomly, but their statistical properties can be determined through testing. *Noise* errors are taken into account both when the range measurements are used in the orbit determination process, and when estimating the accuracy of the orbit determination process.

We shall now assess the effects of station location errors on the range measurement.

If the cartesian co-ordinates of the ground station are (X, Y, Z) with location errors of (dX, dY, dZ) and the spacecraft is at (x, y, z), then the true range ρ is given by:

$$\rho^2 = (x-X)^2 + (y - Y)^2 + (z - Z)^2$$

Differentiating:

$$2\,\rho d\rho = -2\,(x - X)\,dX - 2\,(y - Y)\,dY - 2\,(z - Z)\,dZ$$

$$\therefore \qquad d\rho = \frac{-(x - X)\,dX - (y - Y)\,dY - (z - Z)\,dZ}{\rho}$$

$$\therefore \qquad d\rho = \left(-\frac{(x - X)}{\rho}, -\frac{(y - Y)}{\rho}, -\frac{(z - Z)}{\rho}\right)\begin{pmatrix} dX \\ dY \\ dZ \end{pmatrix}$$

This shows that the effect of the station location errors on the range error depends upon the position of the spacecraft relative to the ground station. The greater the range, the greater is the error.

9.2.1.2 Range Rate

The range rate is the measure of the rate at which the spacecraft to ground station distance is changing. Whereas the range is a measure of radial distance, the range rate is a measure of radial velocity. It is determined from the two-way Doppler shift of the tracking signal frequency. If this is to be measured at the ground station, the spacecraft must employ a *coherent* tracking signal transponder. Although range rate measurements can sometimes usefully augment other measurements such as range and angular position, especially in transfer orbit, their contribution to the overall orbit determination accuracy is usually minor. Furthermore, their use during the geostationary orbit phase is not recommended, as the actual range rate is very small indeed, and measurement errors could degrade the overall orbit determination accuracy.

9.2.1.3 Angular Position

Another type of tracking measurement is the orientation of the spacecraft-to-ground station position vector relative to some station-fixed axis system. Two angles are required as a minimum, and they can be derived either through measurement of the orientation of the tracking antenna or through interferometry. We shall consider each in turn.

Measurement of Antenna Orientation

Antenna pointing angles are measurements of the direction in which the antenna is pointing when making a tracking observation. The measured angles define the line-of-sight to the spacecraft in relation to two rotation axes. The quantities measured depend upon the antenna mount and reference frame, but typical examples are azimuth and elevation angles or right ascension and declination angles. Antenna pointing angles are usually measured in conjunction with ranging, but it is possible to determine the orbit by angle measurements alone. For example, if the spacecraft is equipped with a beacon, instead of a transponder, range measurements are not possible, but the antenna can still track the spacecraft, giving angular measurements of its position.

In the geosynchronous on-station case it is usually preferable for only one station to be responsible for tracking the spacecraft. Range measurements from one station only, however, do not provide good orbit determination accuracy. In this situation, angular measurements are used to supplement the ranging, and together they can achieve the required accuracy. For transfer orbit the spacecraft moves very rapidly near perigee, covering a large arc of the orbit in a short space of time. Consequently, the line-of-sight to the spacecraft may be changing too rapidly for the antenna to follow accurately. Therefore angular measurements, made by a moving antenna, are not usually taken in transfer orbit.

In addition to the angles, the angular rates of the antenna may be measured, but this method is not used very often as it makes only a minor contribution to the overall accuracy.

Interferometry

An interferometer measures the angle between the observer's (tracking station) reference plane and the line from a reference point in the plane, to the spacecraft. See Figure 9.2.

This angle (θ) is obtained by measuring the phase difference between signals arriving at two antennae (A_1 and A_2) which are separated by a known distance (D). Two antennae are not sufficient for the calculation of θ because of the ambiguity arising from D exceeding the wavelength of the signal. To overcome this, additional pairs of antennae are used with different separations in the reference plane.

With this configuration, the measurement of θ is made in one plane only. To locate the spacecraft in 3-dimensional space, a second set of antennae is required, in a plane orthogonal to the first set.

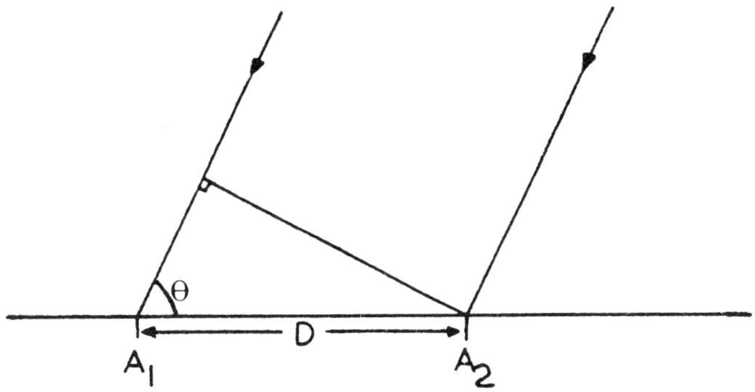

Fig.9.2 Interferometer Geometry

One advantage of interferometers is that the spacecraft does not require a transponder for tracking, instead it can use a low-power beacon which may also serve as a telemetry carrier. This system is not much used for geostationary missions.

9.2.2 Tracking Networks

Several tracking networks are available for the support of geostationary spacecraft. A very large operator of geostationary spacecraft would have its own network, e.g. Intelsat and Telesat, while smaller operators would have recourse to the networks provided by several national and international agencies, e.g. NASA, ESA, and CNES. Within this section we shall briefly discuss the characteristics of these networks.

9.2.2.1 NASA

NASA provides several tracking and data relay networks, but the most important for geostationary missions is the Space Tracking and Data Network (STDN) operating at S-Band frequencies. The network is under the management of the Goddard Space Flight Centre (GSFC), and consists of twelve ground stations, four of which are in the U.S. The network provides world-wide coverage. Each station normally functions in a *store and forward* mode, i.e. tracking data is stored temporarily before being transmitted to GSFC, but when necessary *real time* operation is possible, with the ground stations being connected to GSFC through the *bent-pipe* mode, i.e. using spaceborne links.

Currently, STDN supports about forty spacecraft per day, mostly spacecraft in low Earth orbit.

Table 9.1 gives the salient details of the ground stations comprising the STDN. Figure 9.3 shows the location of the stations.

Table 9.1

Locations of NASA STDN Tracking Stations (Reproduced by kind permission of D. Reidel Publishing Co.)

STATION	SYSTEM	LATITUDE[1]	LONGITUDE (E)	HEIGHT ABOVE ELLIPSOID (METERS)
ASCENSION ISLAND (ACN)	9m USB	– 7°57'17.37''	345°40'22.57''	528
SANTIAGO, CHILE (AGO)	9m USB	–33°09'03.58''	289°20'01 08''	706
	VHF GRARR	–33°09'06.06''	289°20'01.07''	706
	INTERFEROMETER	–33°08'58.10''	289°19'54.20''	694
BERMUDA (BDA)	9m USB	32°21'05.00''	295°20'31.94''	– 33
	FPQ-6 RADAR	32°20'53.05''	295°20'47.90''	– 35
GRAND CANARY ISLAND (CYI)	9m USB	27°45'51.61''	344°21'57 88''	167
ENGINEERING TRAINING CENTER, MARYLAND (ETC)	9m USB	38°59'54.84''	283°09'26.23''	– 1
	9m USB (ERTS)	38°59'54.08''	283°09'29.21''	4
	INTERFEROMETER	38°59'57.25''	283°09'38 71''	– 5
GOLDSTONE, CALIFORNIA (GDS)	26m USB	35°20'29 66''	243°07'35 06''	919
	9m USB (ERTS)	35°20'29 64''	243°07'37 45''	913
GUAM (GWM)	9m USB	13°18'38 25''	144°44'12 53''	116
HAWAII (HAW)	9m USB	22°07'34 46''	200°20'05 43''	1139
	FPS 16 RADAR	22°07'24 37''	200°20'04 02''	1143
MADRID, SPAIN (MAD)	26m USB	40°27'19 67''	355°49'53 59''	808
MERRITT ISLAND, FLORIDA (MIL)	9m USB NO 1	28°30'29 79''	279°18'23 85''	– 55
	9m USB NO 2	28°30'27 91''	279°18'23 85''	– 55
ORRORAL VALLEY AUSTRALIA (ORR)	INTERFEROMETER	–35°37'32 19''	148°57'15 15''	926
QUITO, ECUADOR (QUI)	INTERFEROMETER	–00°37'22 04''	281°25'16 10''	3546
ROSMAN, NORTH CAROLINA (ROS)	4 3m USB	35°11'45 99''	277°07'26 96''	810
	VHF GRARR	35°11'42 02''	277°07'26 97''	810
TANANARIVE, MALAGASY REPUBLIC (TAN)	4 3m USB	–19°01'13 87''	47°18'11 87''	1368
	VHF GRARR	–19°01'16 34''	47°18'11 83''	1368
	INTERFEROMETER	–19°00'31 66''	47°17'59 75''	1347
	FPS 16 RADAR	–19°00'05 52''	47°18'53 46''	1307
FAIRBANKS, ALASKA (ULA)	9m USB	64°58'19 20''	212°29'13 39''	339
	VHF GRARR	64°58'17 50''	212°29'19 12''	339
	INTERFEROMETER	64°58'36 91''	212°28'31 89''	282
WINKFIELD, ENGLAND (WNK)	INTERFEROMETER	51°26'46 12''	359°18'09 13''	87

[1] A MINUS SIGN (–) INDICATES SOUTH LATITUDE.

Note: Quito and Tananarive are no longer available, and Interferometer tracking is discontinued.

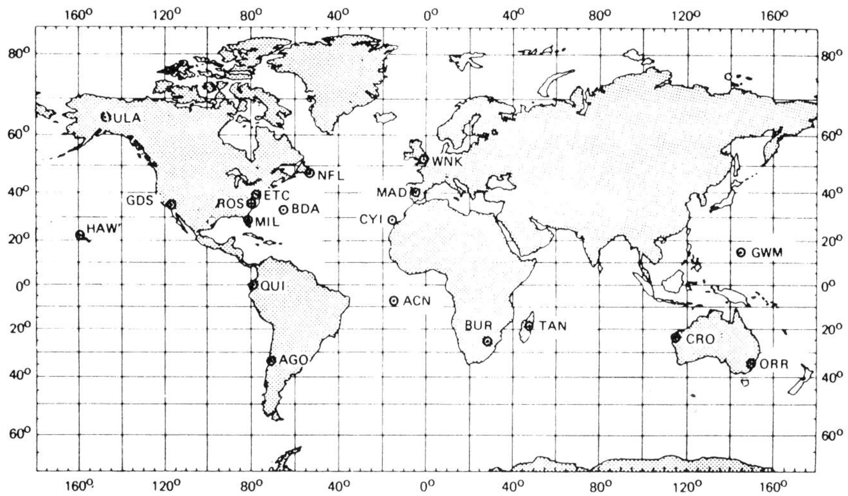

Fig.9.3 Locations of NASA STDN Stations
(The Three-Letter Designators are defined in Table 9.1)
(Reproduced by kind permission of D. Reidel Publishing Co.)

9.2.2.2 ESA

The European Space Agency has its own ground station network to provide TTC services to geostationary spacecraft in their transfer and drift orbit phases. The network is available to users who satisfy the terms of ESA's charter, the most important being that the spacecraft be used for peaceful purposes only.

The network consists of six stations, and is capable of VHF and S–Band operation. VHF will not, however, be retained for very much longer. The network can provide support for two spacecraft simultaneously. Users must specify how many groups of range measurements they require. Each group consists of 32 measurements, and up to 31 groups may be requested.

Table 9.2 shows the location and operating frequencies of the stations within the ESA network. Table 9.3 shows the tones employed within the tone ranging system. Figure 9.4 shows the geographical location of the stations.

Table 9.2

ESA Ground Stations

Ground Station	Long. E. of GM		Latitude, N		Freq. Band
	Deg.	Min.	Deg.	Min.	
Kourou/Galliot (French Guyana)	307	12	5	15	UHF
Kourou/Diane (French Guyana)	307	12	5	15	VHF
Malindi (Kenya)	40	19	-3	0	UHF & VHF
Villafranca (Spain)	-3	57	40	27	UHF
Carnarvon (W. Australia)	113	43	-24	54	UHF & VHF
Redu (Belgium)	5	8	50	0	VHF

Table 9.3

Ranging Tones of the ESA Ground Stations

Ranging Band	Major Tone (kHz)	Minor Tones (kHz)
UHF	100	20.0 16.0 16.8 16.16 16.032 16.008
VHF	20	15.0 16.250 15.250 15.050 15.010

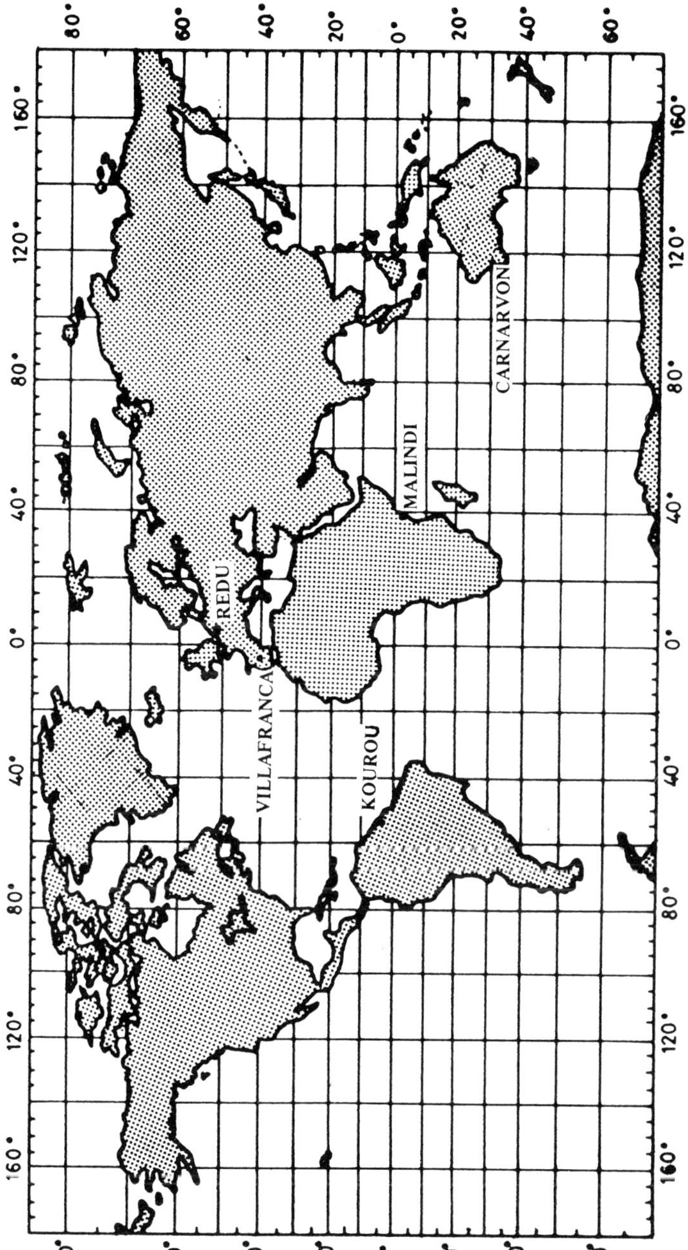

Fig.9.4 Geographical Location of ESA Ground Stations

9.2.2.3 Intelsat/Comsat

The Comsat ground station network is used for the transfer and drift orbit services to Intelsat spacecraft. The network can operate at either C–Band or Ku–Band. A tone–ranging system is used employing one of two alternative tone groups. The network consists of six stations providing global coverage. Table 9.4 indicates the positions of the stations of the network. They are controlled from the control centre located in Washington D.C., USA. The network can achieve a range resolution, subsequent to all processing, of better than 15 metres, for a received signal–to–noise ratio of at least 20 dB.

9.2.2.4 Telesat

The Telesat network supports the Anik series of spacecraft providing communications services to Canada. The network consists of two strategically located stations at Allan Park (near Toronto) and Perth (W. Australia). The network operates at C–Band, and a Ku–Band capability is also available. Both the Allan Park and the Perth antennae can be driven along the satellite's predicted path under computer control. In addition, a computer–generated search pattern can be super–imposed on the track.

Station characteristics are outlined in Table 9.5.

9.2.2.5 CNES

The CNES ground station network is used primarily for control of geostationary spacecraft through all phases of the mission. The network consists of the control centre at Toulouse, France, and three stations: Kourou, Hartebeeshoek, and Aussaguel. The locations of these stations is given in Table 9.6. In addition to these stations, arrangements have been made for two NASA stations, Canberra and Goldstone, to provide additional support when required. The network operates at S–Band. The system is configured for high reliability – the data links between the out–stations and the control centre are duplicated, as are the computers in the control centre. The availability of the network to support spacecraft operations is 99.7%. The probability of a successful outcome to the control activities is 99.5%.

For tracking, range and range–rate measurements are possible. The antenna can track the spacecraft in three modes:

- manual;
- auto–track on the spacecraft telemetry beacon;
- use of pointing data (program tracking).

9.2.2.6 U.S. Air Force Satellite Control Facility (AFSCF)

The AFSCF's Space–Ground Link Subsystem (SGLS) provides for the tracking of orbiting military spacecraft. The AFSCF has twelve tracking stations at seven geographic locations controlled from the satellite Test Centre (STC) at Sunnyvale,

California – see Figure 9.5. The network operates at S-Band, and tracking measurements include range, range-rate, and antenna angles.

The ranging system employs a pseudo-random noise coded signal. The transmitted and received signals are correlated, and the time delay yielding maximum correlation is extracted. If this delay is T, the range is TC/2, where C is the velocity of wave propagation.

Table 9.7 shows the range and angle accuracies possible with the system. Table 9.8 outlines the characteristics of the stations comprising the network.

Table 9.4

Location of COMSAT Ground Stations

Ground Station	Longitude (deg. E. of GM)	Latitude (deg. N)
Southbury (Connecticut, U.S.)	286.71	41.45
Santa Paula (California, U.S.)	240.93	34.40
Paumalu (Hawaii)	201.97	21.67
Seychelles	–4.66	55.3
Guam	144.83	13.5
Andover (New Hampshire, U.S.)	288.34	40.8

Table 9.5

Characteristics of TELESAT Stations

	Allan Park (Canada)	Perth (W. Australia)
Band	C	C and Ku
Antenna Diameter	11 m	8 and 10 m
Shaft Encoder Resolution	0.001 deg.	0.001 deg.
Tracking Mode	3 channel, monopulse	Monopulse
Tracking Velocity	1.0 deg./sec.	1.0 deg./sec.
Autotrack Error	0.007 deg. (rms)	0.01 deg. (rms)

Table 9.6

Locations of CNES Network Stations

Location	Network	Longitude (° East)	Latitude (° North)	Altitude (m)
Aussaguel (France)	CNES	1° 26'	43° 30'	210
Hartebeeshoek (S. Africa)	CNES	28° 22'	-25° 33'	1191
Kourou (French Guyana)	CNES	307° 12'	5° 15'	113
Goldstone (U.S.A.)	NASA	243° 07'	35° 20'	915
Canberra (Australia)	NASA	148° 57'	-35° 37'	933

Telemetry Reception Frequency: 2200–2290 MHz
G/T = 22 dB/°k
Minimum flux density = –182 dBW/m^2

Telecommand Frequency: 2025–2110 MHz
Antenna Maximum EIRP = 71 dBW.

Table 9.7

Measurement Accuracy Standard Deviations for the SGLS Stations

Measurement	Noise Error	Bias Error
Range (m)	10	8.0
Azimuth (deg)	0.017	0.017
Elevation (deg)	0.017	0.017
Station location (each axis)	10m	

Note: (a) The range bias error assumes that ionospheric and tropospheric bias errors have been removed.

(b) The angle measurement errors are for a geostationary orbit.

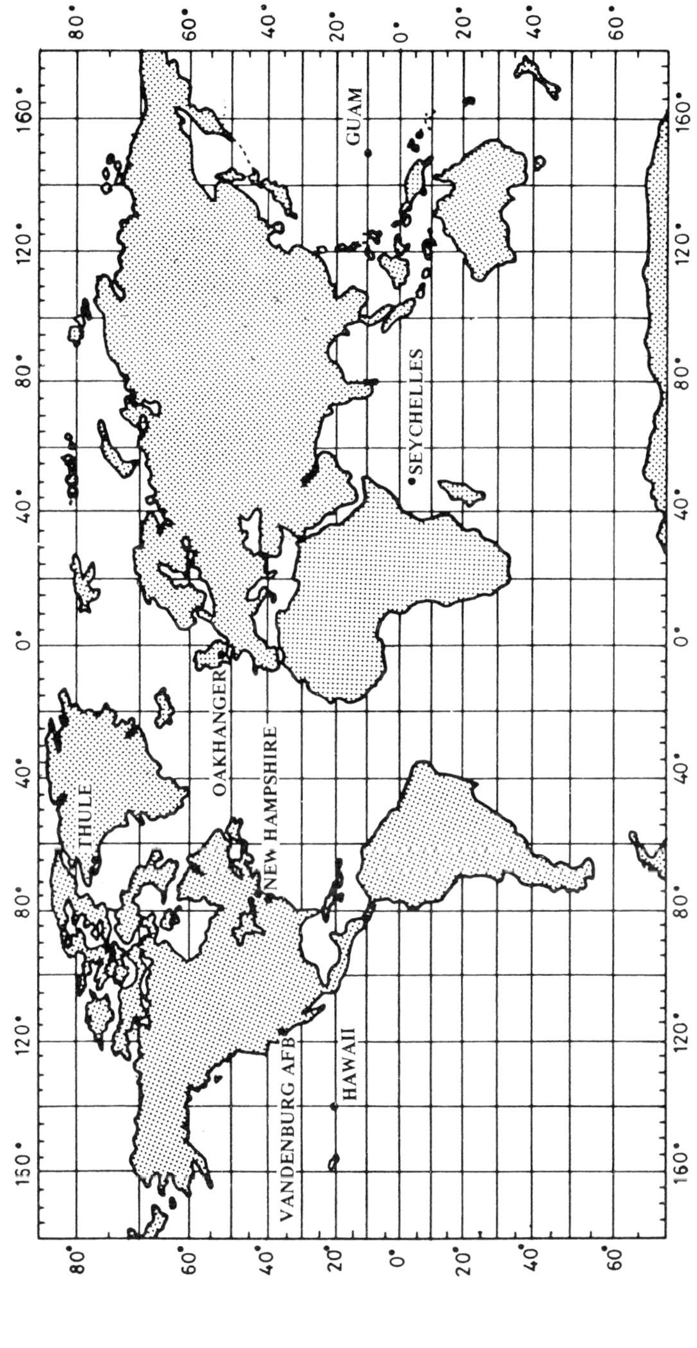

Fig.9.5 Geographical Location of SGLS Ground Stations

Table 9.8

AFSCF Ground Station Details

Tracking Station	Latitude N deg	min	Longitude deg	min	Altitude m	Acquire & autotrack i.e. 400 MHz band	Can only operate in 2.2 GHz band
NHS – New Hampshire							
TTC 60 ft	42	56.9	288	22.4	200.8	√	
TTC 46 ft	42	56.7	288	22.4	190.7		√
VTS – Vandenburg AFB (California)							
TTC 60 ft	34	49.4	239	29.9	269.4	√	
TTC 46 ft	34	49.6	239	29.7	265.5		√
HTS Hawaii (Kaena Point, Oahu)							
TTC 60 ft	21	33.8	201	45.5	428.2	√	
TTC 46 ft	21	34.1	201	44.3	317.9		√
GTS – Guam							
TTC 60 ft	13	36.9	144	52.0	218.4		√
TTC 46 ft	13	36.95	144	51.3	208.8		√
IOS – Indian Ocean (Mahi Island, Seychelles) TTC 60 ft	–4	40.3	55	28.7	561.4	√	
TTS – Thule (Greenland)							
TTC 46 ft	76	30.9	291	24.0	132.0		√
TTC 14 ft	76	31.0	291	24.0	131.9		√
OTS – Oakhanger (England)							
WAT 60 ft	51	6.8	359	6.3	–21.9		√

TTC Tracking Telemetry and Command
WAT Wheel and Track
Altitude with respect to reference geoid.

9.3 Orbit Determination

Orbit determination establishes the position of the spacecraft in a suitable reference frame, and the characteristic elements of its orbit. It can function in two basic modes, orbit refinement of a previously established orbit, and orbit determination without any prior knowledge of the orbital elements. During a spacecraft's mission it is the former that is almost always carried out because some estimate of the spacecraft's orbit is usually available. We shall, however, discuss both approaches. In the subsequent developments vector quantities will be represented in bold type.

9.3.1 First Acquisition Orbit Determination

To illustrate the method we shall consider the simplest case where a ground station makes measurements of a spacecraft's position vector at two times, less than an orbit period apart.

Let $\mathbf{R}_1$ be the position vector of a spacecraft as measured from the ground station at time t_1, and $\mathbf{R}_2$ the position vector at a later time t_2. There are infinitely many orbits passing through $\mathbf{R}_1$ and $\mathbf{R}_2$, but only one having the measured time of flight $T = t_2 - t_1$. Hence the measurements $\mathbf{R}_1$, $\mathbf{R}_2$, t_1 and t_2 uniquely determine the orbit. We now convert $\mathbf{R}_1$ and $\mathbf{R}_2$ to spacecraft position vectors in the geocentric inertial reference frame, and call these $\mathbf{r}_1$ and $\mathbf{r}_2$. The conversion consists of sequential axis rotations, taking into account the rotation of the Earth between times t_1 and t_2.

Hence the measurements $\mathbf{r}_1$, $\mathbf{r}_2$, t_1 and t_2 now uniquely define the orbit. In addition the change in the true anomaly, Δf, between t_1 and t_2 is known, since

$$\mathbf{r}_1 = r_1 . \mathbf{u}_1 \quad , \quad \mathbf{r}_2 = r_2 . \mathbf{u}_2 \quad ,$$

and $\quad \cos \Delta f = \mathbf{u}_1 . \mathbf{u}_2$

We now make the assumption that:

$$\mathbf{r}(t_2) = F\, \mathbf{r}(t_1) + G\, \mathbf{v}(t_1). \tag{1}$$

This states that $\mathbf{r}(t_2)$ may be expressed in terms of $\mathbf{r}(t_1)$, and $\mathbf{v}(t_1)$, where $\mathbf{v}(t_1)$ is the velocity vector of the spacecraft whose position vector is $\mathbf{r}(t_1)$. Now, $[\mathbf{r}(t_1), \mathbf{v}(t_1)]$ is the state vector of the spacecraft, which uniquely defines its orbit. If the time of flight, $(t_2 - t_1)$, is less than the orbital period, then it is reasonable to assume that $\mathbf{r}(t_2)$ lies in the plane containing the vectors $\mathbf{r}(t_1)$ and $\mathbf{v}(t_1)$. The orbit geometry is shown in Figure 9.6.

We shall simplify the notation by writing:

$$\mathbf{r}_2 = F\, \mathbf{r}_1 + G\, \mathbf{v}_1$$

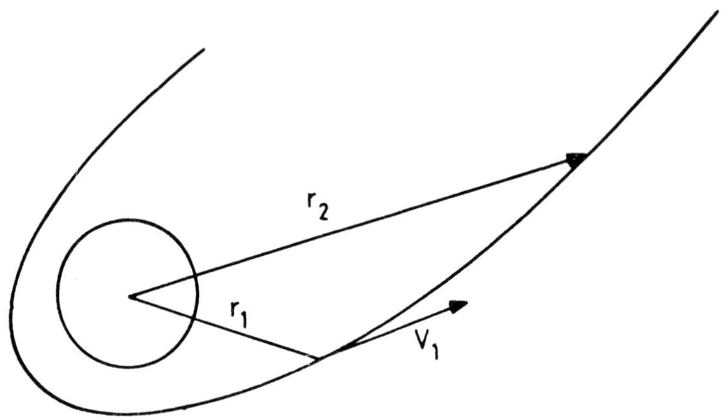

Fig.9.6 Orbit Geometry

It can be shown that the F and G coefficients can be expressed in closed form as:

$$F = 1 - \frac{r_2}{p}(1 - \cos \Delta f) = 1 - \frac{a}{r_1}(1 - \cos \Delta E) \tag{2}$$

and

$$G = \frac{r_1 r_2 \sin \Delta f}{\sqrt{\mu p}} = T - \sqrt{\frac{a^3}{\mu}}(\Delta E - \sin \Delta E) \tag{3}$$

where:
r_1 = $|\mathbf{r}_1|$,
r_2 = $|\mathbf{r}_2|$,
p = semi-latus rectum = $a(1-e^2)$,
a = semi-major axis,
e = eccentricity,
E = eccentric anomaly,
f = true anomaly,
ΔE = $E_2 - E_1$,
Δf = $f_2 - f_1$,
μ = gravitational constant.

Determination of the F and G coefficients in Eqn. (1) enables one to determine the velocity vector at time t_1 through the expression:

$$\mathbf{v}_1 = (\mathbf{r}_2 - F\,\mathbf{r}_1)/G$$

Once the spacecraft state vector, $(\mathbf{r}_1, \mathbf{v}_1)$, at time t_1 is known, the osculating orbit elements at this epoch can be determined. These elements can then be used in conjunction with an appropriate force model to predict the spacecraft's future motion. This then facilitates re-acquisition of the spacecraft at some later time for more tracking observations, and also provides a first estimate of the orbit elements for use with orbit refinement techniques discussed later.

Let us now proceed with the derivation of the orbit. After tracking, $\mathbf{r}_1$, $\mathbf{r}_2$, Δf, and T are known. The unknowns in Eqns. (2) and (3) are a, e, and ΔE. The orbit elements i, Ω, and ω, which are the inclination, right ascension of the ascending node, and argument of perigee respectively, can be determined once the state vector is defined. Eqns. (2) and (3) are transcendental in nature, so an iterative method of solution is adopted. The following procedure is typical.

To start the process we choose an initial value of p.

$$p^{(o)} = (r_1 + r_2)/2$$

This is merely an estimate and any other value which is considered to be a good estimate would suffice.
Since:

$$r = p \,/\, (1 + e \cos f)$$

$$e \cos f_i = \left(\frac{p^{(o)}}{r_i}\right) - 1 \qquad \text{for } i = 1 \text{ and } 2 \tag{4}$$

Hence:

$$e \sin f_1 = [\cos \Delta f\,(e \cos f_1) - (e \cos f_2)]/\sin \Delta f \tag{5}$$

and

$$e \sin f_2 = [-\cos \Delta f\,(e \cos f_2) + (e \cos f_1)]/\sin \Delta f \tag{6}$$

We then have:

$$e = [(e \cos f_1)^2 + (e \sin f_1)^2]^{1/2}$$

$$a = p^{(o)}/(1 - e^2) \tag{7}$$

$$n = \sqrt{\frac{\mu}{a^3}}$$

The eccentric anomalies are then given by:

$$\sin E_i = \frac{r_i \sin f_i}{a\,(1 - e^2)^{1/2}} \qquad \text{for } i = 1 \text{ and } 2$$

$$\cos E_i = \frac{r_i}{a} \cos f_i \qquad \text{for } i = 1 \text{ and } 2 \tag{8}$$

We shall now calculate the time it takes the spacecraft to travel from $\mathbf{r}_1$ to $\mathbf{r}_2$ on the estimated orbit, and compare this with the observed time of flight, $t_2 - t_1 = T$. The time of flight along the estimated orbit is given by:

$$T^{(o)} = \frac{1}{n}\,(M_2 - M_1) \tag{9}$$

where M_1 and M_2 are the mean anomalies at times t_1 and t_2. From Kepler's equation we have:

$$T^{(o)} = \frac{1}{n}\,[(E_2 - E_1) + e\,(\sin E_1 - \sin E_2)] \tag{10}$$

To compare this time of flight with the observed one, we form the function H given by:

$$H = T - T^{(o)} \tag{11}$$

If $H \neq 0$, we choose a new value of $p = p^{(1)}$, and recalculate H using the Eqns. (4) to (11). This process is continued iteratively until $|H| < \epsilon$, where ϵ is some suitably chosen small tolerance.

The choice of the new value of p is an important part of the orbit determination technique, and must ensure rapid and stable convergence. Let $p^{(j)}$ be the current value of p, then the next value of p, $p^{(j+1)}$, is calculated as follows.

We increment $p^{(j)}$ by Δp, and use Eqns. (4) to (11) to evaluate:

$$H = H\,(p^{(j)} + \Delta p)$$

An increment of approximately 5% ($\Delta p / p^{(j)} = 0.05$) is usually adopted. We can then estimate the derivative of H with respect to p:

$$H'(p^{(j)}) = \frac{H\,(p^{(j)} + \Delta p) - H\,(p^{(j)})}{\Delta p}$$

The new value of $p = p^{(j+1)}$ is then obtained by Newton's Method:

$$p^{(j+1)} = p^{(j)} + \frac{H(p^{(j)})}{H'(p^{(j)})}$$

9.3.2 Orbit Refinement

Orbit refinement is the most commonly used technique of orbit determination, because an initial estimate of the orbit is usually available. For example, an estimate of the transfer orbit is provided by knowledge of the launch vehicle's trajectory; an estimate of the drift orbit is obtained by knowledge of the transfer orbit and the apogee manoeuvre. With this technique, orbit observations are used to improve knowledge of the orbit elements at some epoch.

Two basically different orbit refinement methods are available: the *batch processing* method and the *sequential filter* method. Both have their advantages and disadvantages, which we shall discuss next.

The *batch processing* method consists of making observations over a time interval, and processing them to obtain a better estimate of the orbit at some epoch *within* the time interval. It is ensured that the number of observations greatly exceeds the number of unknowns to allow a *least squares* estimation process to be used. The advantage of this method is that the procedure can easily be made highly accurate and numerically stable. A disadvantage is that the estimate of the orbit can only be improved step-wise, and infrequently, owing to the need to collect observations over a certain time interval (usually of the order of hours).

The *sequential filtering* method consists of using the observations *as they are made*, and obtaining a progressively improved estimation of the orbit at the time of the most recent observation. The filtering technique used is that of the *extended Kalman filter*. The advantage of this method is that the improved estimate is obtained much more quickly, and the orbit refinement process is a continuous rather than a step-wise one. The disadvantage is that the procedure can lead to numerical instability over extended durations.

Below we shall discuss each method in turn.

9.3.2.1 The Weighted Least-Squares Method

Let the initial estimate of the orbit parameters at epoch be given by the vector $\epsilon^{(o)}$ defined by:

$$\epsilon^{(o)}_{(N\times 1)} = \begin{bmatrix} \epsilon_1 \\ \vdots \\ \epsilon_N \end{bmatrix}$$

where N is the total number of parameters to be refined. Though the method applies to any consistent parameter set, the choice of parameters depends strongly upon the type of orbit being considered.

For an elliptic inclined Earth orbit such as a geostationary transfer orbit, the parameter set may be defined as:

$$\epsilon^{(o)^T}_{(1\text{x}8)} = (a \; e \; i \; \Omega \; \omega \; M_0 \; M_1 \; M_2)$$

where a, e, i, Ω, and ω are the usual Keplerian elements, and the mean anomaly is given by:

$$M(t) = M_0 + M_1 t + M_2 t^2$$

The coefficients M_1 and M_2 are useful in modelling the along-track deceleration of the spacecraft owing to the variation in aerodynamic drag during the tracking interval.

For a geostationary orbit, we may employ the more useful set:

$$\epsilon^{(o)^T}_{(1\text{x}6)} = (\lambda_s \; \dot{\lambda}_s \; e_x \; e_y \; i_x \; i_y)$$

where:
λ_s = longitude of the satellite,
$\dot{\lambda}_s$ = longitude drift rate,
e_x = X-component of the eccentricity vector,
e_y = Y-component of the eccentricity vector,
i_x = X-component of the orbit pole vector,
i_y = Y-component of the orbit pole vector.

If the amount of tracking data is large we may also include other parameters such as solar radiation pressure and the magnitude of the spacecraft's orbital manoeuvres. Alternatively, the parameter set $\epsilon^{(o)}$ could be the state vector at epoch:

$$\epsilon^{(o)^T} = (\mathbf{r}, \mathbf{v}) \text{ at epoch}$$

The satellite is tracked over a period of time spanning the epoch. For the tracking of a spacecraft in geostationary orbit, this period could be one or two days. The tracking campaign results in the generation of k observed quantities, η_k^{ob}, $k = 1 \ldots\ldots K$, at times $t = t_1 \ldots\ldots t_K$. More than one measurement may be made concurrently. For example, if the range, ρ, azimuth, θ, and elevation, β, are measured at time $t = t^1$ then:

$$\eta_i^{ob} = \rho(t_i) = \rho(t^1),$$

$$\eta_{i+1}^{ob} = \theta(t_{i+1}) = \theta(t^1),$$

$$\eta_{i+2}^{ob} = \beta(t_{i+2}) = \beta(t^1).$$

The spacecraft position at the time of each observation $t = t_1$ ..... t_K is then evaluated by propagating the orbit with a suitable force model, using the initial parameter set $\epsilon^{(o)}$. Simulated tracking measurements are now computed, η_k^{comp}, $k = 1$ ..... K. Since the actual spacecraft position will differ from the simulated one based on $\epsilon^{(o)}$, this procedure gives the observed-minus-computed observation residuals:

$$\Delta\eta_k^{(o)} = \eta_k^{ob} - \eta_k^{comp}, \quad k = 1 \;.....\; K$$

A measure of the fit of the observations to the initial estimate of the orbit is given by the sum of the squares of the weighted measurement residuals.

We have, therefore

$$\psi^{(o)} = \sum_{k=1}^{K} \frac{1}{\sigma_k^2} (\Delta\eta_k^{(o)})^2 = \Delta\eta^{(o)^T} \mathbf{W}^{-1} \Delta\eta^{(o)}$$

where $\mathbf{W}^{-1}$ is the diagonal weighting matrix reflecting the accuracy of the observations:

$$\mathbf{W}^{-1}(K \times K) = \begin{bmatrix} \frac{1}{\sigma_1^2} & & 0 \\ & \ddots & \\ 0 & & \frac{1}{\sigma_K^2} \end{bmatrix}$$

Here σ_k^2 is the variance of the observed quantity η_k^{ob}. The purpose of including the weighting matrix in the calculation of ψ is to place greater reliance on more accurate observations.

The observation residuals are functions of the parameter set components thus:

$$\Delta\eta_k^{(o)} = \sum_{j=1}^{N} \frac{\delta\eta_k}{\delta\epsilon_j} \Delta\epsilon_j \qquad k = 1 \;.....\; K$$

In matrix form this can be written:

$$\Delta\eta^{(o)}_{(k \times 1)} = \mathbf{A}_{(k \times N)} \; \Delta\epsilon_{(N \times 1)} \tag{1}$$

where:

$$\Delta\eta^{(o)^T} = (\Delta\eta_1^{(o)} \;.....\; \Delta\eta_K^{(o)})$$

$$\Delta\epsilon^T \doteq (\Delta\epsilon_1 \;.....\; \Delta\epsilon_N)$$

and:

$$\mathbf{A} = \begin{bmatrix} \frac{\delta\eta_1}{\delta\epsilon_1} & \cdots\cdots & \frac{\delta\eta_1}{\delta\epsilon_N} \\ \vdots & & \vdots \\ \frac{\delta\eta_K}{\delta\epsilon_1} & \cdots\cdots & \frac{\delta\eta_K}{\delta\epsilon_N} \end{bmatrix}$$

The elements of **A** must be determined, either analytically or numerically. Analytic derivations are usually possible and result in a saving of computer time. Since there are errors in the observations and in the force model used to propagate the simulated spacecraft position, Eqn. (1) should be restated as:

$$\overset{(o)}{\Delta\eta}_{(K\times 1)} = \mathbf{A}_{(K\times N)}\,\Delta\epsilon_{(N\times 1)} + \mathbf{E}_{(K\times 1)} \tag{2}$$

where **E** is a vector containing the error in the residuals due to the above-mentioned sources.

If K is less than N, the system is indeterminate. If K is greater than N it is over-determined, and a least-squares solution is possible. We require to minimise the sum of the squares of the weighted errors, that is, we minimise the scalar quantity ϕ_E, where:

$$\phi_E = \mathbf{E}^{\mathrm{T}}_{(1\times K)}\,\mathbf{W}^{-1}_{(K\times K)}\,\mathbf{E}_{(K\times 1)}$$

Hence we require that:

$$\frac{\delta\phi_E}{\delta(\Delta\epsilon)} = \mathbf{0}_{(N\times 1)}$$

Using Eqn. (2) we can write ϕ_E as:

$$\phi_E = (\overset{(o)}{\Delta\eta} - \Delta\epsilon^{\mathrm{T}}\,\mathbf{A}^{\mathrm{T}})\,\mathbf{W}^{-1}\,(\overset{(o)}{\Delta\eta} - \mathbf{A}\,\Delta\epsilon)$$

We must now form the differential, $\delta\phi_E/\delta(\Delta\epsilon)$, by noting that:

$$\frac{\delta(\mathrm{y}^{\mathrm{T}}\,\mathrm{x})}{\delta\mathrm{x}} = \mathrm{y} \text{ and } \frac{\delta(\mathrm{x}^{\mathrm{T}}\,\mathrm{y})}{\delta\mathrm{x}} = \mathrm{y}$$

Then:

$$\frac{\delta\phi_E}{\delta(\Delta\epsilon)} = -2\,(\mathbf{A}^{\mathrm{T}}\,\mathbf{W}^{-1}\,\overset{(o)}{\Delta\eta}) + 2\,(\mathbf{A}^{\mathrm{T}}\,\mathbf{W}^{-1}\,\mathbf{A}\,\Delta\epsilon) = \mathbf{0}$$

Hence:

$$\Delta\epsilon = (A^T W^{-1} A)^{-1} (A^T W^{-1} \Delta\eta^{(o)})$$

The solution of Eqn. (3) provides the new parameter set:

$$\epsilon^{(1)} = \epsilon^{(o)} + \Delta\epsilon$$

which is a refinement of the initial estimate $\epsilon^{(o)}$ at epoch. With the new parameter set, $\epsilon^{(1)}$, new simulated observations can be calculated, η_k^{comp}, $k = 1 .....K$. These lead to new observation residuals, $\Delta\eta^{(1)}$, which in turn gives:

$$\psi^{(1)} = \Delta\eta^{(1)^T} W^{-1} \Delta\eta^{(1)}$$

If the parameter set $\epsilon^{(1)}$ is an improvement over $\epsilon^{(o)}$, then $\psi^{(1)} < \psi^{(o)}$.

It can be clearly seen that the procedure for obtaining a final parameter set at epoch is an iterative one – given a parameter set $\epsilon^{(j)}$, observation residuals $\Delta\eta^{(j)}$ are calculated, and $\psi^{(j)}$ evaluated by:

$$\psi^{(j)} = \Delta\eta^{(j)^T} W^{-1} \Delta\eta^{(j)}$$

The increment vector $\Delta\epsilon$ is then obtained by a least–squares solution, to yield a new parameter set:

$$\epsilon^{(j+1)} = \epsilon^{(j)} + \Delta\epsilon$$

Observation residuals, $\Delta\eta^{(j+1)}$, based upon $\epsilon^{(j+1)}$ are computed to give:

$$\psi^{(j+1)} = \Delta\eta^{(j+1)} W^{-1} \Delta\eta^{(j+1)}$$

The convergence of the procedure is monitored by calculating δ at each step, where:

$$\delta^{(j+1)} = (\psi^{(j)} - \psi^{(j+1)})/\psi^{(j)}$$

The value of δ must decrease as the iteration proceeds. The iteration stops when successive values of δ are insufficiently different from each other. Convergence is not guaranteed however, and if δ increases, the iteration must be stopped. Common reasons for lack of convergence are a poor initial orbit parameter set, $\epsilon^{(o)}$, or poor observations. Another reason could be that the tracking interval is too long, and allows propagation errors to develop owing to inadequacies in the orbit force model.

Note that the elements of the matrix **A**, a_{ij}, express how much the observation η_i^{ob} is expected to change if the orbit parameter, ϵ_j is varied. A solution for ϵ_j is possible only if its value has a measurable effect on at least one η_i. The parameter ϵ_j is then said to be *observable*. Good observability occurs when the change in η^{ob} is of the same order as, or greater than, the change in

ϵ_j. Poor observability occurs when the change in η^{ob} is much smaller than the change in ϵ_j. Tracking is much more accurate and effective if the observation types employed offer good observability of the orbit parameters. If there is no observability of one or more of the parameters ϵ_j, the matrix $(\mathbf{A}^T \mathbf{W}^{-1} \mathbf{A})^{-1}$ becomes singular, and no solution is possible. The degree of observability will be reflected in the nature of the matrix $(\mathbf{A}^T \mathbf{W}^{-1} \mathbf{A})^{-1}$. A high degree of observability gives a well-conditioned matrix, while a poor degree of observability gives a poorly conditioned matrix, with some elements very much larger in absolute terms, than others. A poorly conditioned matrix can cause numerical problems in the computation, and even divergence of the iteration.

The residuals corresponding to a particular measurement type, e.g. range, from a particular ground station, can be examined. If the residuals are randomly distributed with zero mean value, the orbit determination accuracy cannot be further improved. If the mean value of the residuals is not zero, it is possible that there is a bias in the measurement system or in the orbit propagation model. We can then improve the orbit determination accuracy by including the measurement bias in the solution vector, ϵ. Once the value of the bias is known, it can be removed from the measurement. Expansion of the solution vector, ϵ, by the inclusion of measurement biases requires an abundance of tracking data, so that the least squares approach can be applied.

9.3.2.2 The Extended Kalman Filter Method

The Kalman filter method provides an estimate of the *state* of a dynamical system from *noisy* observations of the system. When the method is used for orbit refinement, the state vector consists of the orbital elements and the measurements are the spacecraft-to-ground station, range, range-rate, angles, etc. The filter provides a *real time* or *sequential* estimate of the *current* orbit elements by processing the observations as they become available. In addition, it provides an estimate of the error covariance of the orbit elements. The *classical* Kalman filter can only be applied to *linear* dynamical systems. The orbit refinement problem, however, is a non-linear one both in terms of the dynamics of the system and the measurements. The method has, therefore, been extended to incorporate non-linearities into the procedure.

The dynamical system is governed by the *state* equation:

$$\frac{d\epsilon}{dt} = \mathbf{f}\ (\epsilon(t), t) \tag{1}$$

and the *observation* equation:

$$\eta = \mathbf{h}\ (\epsilon(t), t)$$

where: ϵ is the state vector (the orbit element vector),
η is the observation vector.

Let us assume, for the present, that the functions **f** and **h** are known exactly, and that the observations are without error. This implies that we have a perfect *model* of the dynamical system. Let us further assume that an *a priori* estimate of the state vector, ϵ^O, is available. This must satisfy Eqn. (1). The deviation from the *a priori* estimate is, therefore, given by:

$$\Delta\epsilon = \epsilon - \epsilon^O.$$

Similarly, the observation vector corresponding to the *a priori* estimate of the state vector, ϵ^O, is given by:

$$\eta^O = \mathbf{h}\,(\epsilon^O(t), t)$$

The deviation between this and the actual observation vector is given by:

$$\Delta\eta = \eta - \eta^O$$

We now express the function, **f**, in terms of the *a priori* estimate of the state vector by expanding the function in a Taylor series about the *a priori* estimate ϵ^O:

$$\mathbf{f}\,(\epsilon(t), t) = \mathbf{f}\,(\epsilon^O(t), t) + \frac{\delta\mathbf{f}\,(\epsilon^O\,(t), t)}{\delta\epsilon^T}\,.\,\Delta\epsilon + \ldots.. \tag{2}$$

Quadratic and high-order terms are ignored, and Eqn. (2) is substituted into Eqn. (1) to yield:

$$\frac{d\,(\Delta\epsilon + \epsilon^O)}{dt} = \mathbf{f}\,(\epsilon^O(t), t) + \frac{\delta\mathbf{f}\,(\epsilon^O(t), t)}{\delta\epsilon^T}\,.\,\Delta\epsilon$$

But: $$\frac{dc^O}{dt} = \mathbf{f}\,(\epsilon^O(t), t)$$

Therefore: $$\frac{d(\Delta\epsilon)}{dt} = \mathbf{F}\,(\epsilon^O(t), t).\,\Delta\epsilon \tag{3}$$

where: $$\mathbf{F}(\epsilon^O(t), t) = \frac{\delta\mathbf{f}\,(\epsilon^O(t), t)}{\delta\epsilon^T} \tag{3a}$$

is the *Jacobian* matrix.

Eqn. (3) is a linearised state equation relating the *deviation* of the state vector from its *a priori* estimate. Using the same approach for the observation function, h, we have:

$$\mathbf{h}\,(\epsilon(t), t) = \mathbf{h}\,(\epsilon^O(t), t) + \frac{\delta\mathbf{h}\,(\epsilon^O(t), t)}{\delta\epsilon^T}\quad\Delta\epsilon + \ldots..$$

This gives:

$$\Delta\eta = \mathbf{H}(\epsilon^{\mathrm{o}}(t), t).\,\Delta\epsilon \tag{4}$$

where the observation matrix, **H**, is given by:

$$\mathbf{H} = \frac{\delta\mathbf{h}(\epsilon^{\mathrm{o}}(t), t)}{\delta\epsilon^{\mathrm{T}}} \tag{4a}$$

Eqns. (3) and (4) describe a linear system in *state*, and *observation*, *deviations*, governed by a differential equation set. To mechanise the procedure for a digital computer, we must express the relationships as difference equations at discrete points in time, related to the times at which the observations are made. Let us discretise time by the sequence of points (t_k). We have therefore:

$$\Delta\epsilon_{k+1} = \phi(t_{k+1}, t_k)\,\Delta\epsilon_k \tag{5}$$

where $\phi(t_{k+1}, t_k)$ is the state transition matrix which transforms the state deviations at time t_k to those at time t_{k+1}. The state transition matrix obeys the following relationships:

$$\frac{d\phi(t,\tau)}{dt} = \mathbf{F}(t)\,\phi(t,\tau) \text{ for all } t \text{ and } \tau\ ,$$

$$\phi(\tau,\tau) = \mathbf{I} \text{ (the \textit{identity} matrix) for all } \tau\ ,$$

$$\phi(\tau,t)\,.\,\phi(\lambda,\tau) = \phi(\lambda,t) \text{ for all } t, \tau, \text{ and } \lambda.$$

The state transition matrix may be obtained from the relation:

$$\phi(t_{k+1}, t_k) = \exp\,[(t_{k+1} - t_k)\,\mathbf{F}(\epsilon^{\mathrm{o}}(t_k), t_k)] \tag{6}$$

The difference equation of the observation deviation is:

$$\Delta\eta_k = \mathbf{H}(\epsilon^{\mathrm{o}}(t_k), t_k)\,\Delta\epsilon_k \tag{7}$$

Eqns. (5), (6), and (7) describe a discretised linear system.
Let us now review the assumptions made earlier. These were:

- the dynamical model is perfect,
- the observations are perfect.

In practice neither assumption is tenable. Not only is the non-linear model imperfectly known, but it is also linearised. The discretisation process introduces futher errors. Finally, there is some error associated with the observations. Let us assume that all these errors result in *noisy* estimates of the state, and observation, deviations. We can, therefore, express Eqns. (5) and (7) as:

$$\Delta\epsilon_{k+1} = \phi\,(t_{k+1}, t_k)\,.\,\Delta\epsilon_k + \mathbf{W}_{k+1}$$

$$\Delta\eta_{k+1} = \mathbf{H}\,(\epsilon^{o}(t_k), t_k).\,\Delta\epsilon_k + \mathbf{V}_{k+1}$$

where $\mathbf{W}_{k+1}$ and $\mathbf{V}_{k+1}$ are the random noise vectors associated with the state and observation vectors respectively. Let us assume that the noise vectors behave as *Gaussian white noise*. Then:

$$E\,\{\mathbf{W}_k\} = E\,\{\mathbf{V}_k\} = 0$$

$$\mathrm{E}\,\{\mathbf{W}_k\,\mathbf{W}_j\} = \mathbf{Q}_k\,\delta_{kj}$$

$$E\,\{\mathbf{V}_k\,\mathbf{V}_j\} = \mathbf{R}_k\,\delta_{kj}$$

where $E\,\{\mathbf{x}\}$ denotes the expectation of $\mathbf{x}$, $\mathbf{Q}_k$ is the state noise covariance matrix, $\mathbf{R}_k$ is the observation noise covariance matrix, and δ_{kj} is the Kronecker delta. Finally, we assume that the state and observation noise vectors are uncorrelated:

$$E\,\{\mathbf{V}_k\,\mathbf{W}_j^{\mathrm{T}}\} = 0 \text{ for all } k \text{ and } j.$$

The procedure of the extended Kalman filter uses the best estimate of the present step as the *a priori* estimate for the next. At each step the linearised estimates for ϕ and $\mathbf{H}$ are derived using Eqns. (3a), (4a), and (6). The original non-linear equations:

$$\frac{d\epsilon}{dt} = \mathbf{f}\,(\epsilon(t), t) \text{ and } \eta = \mathbf{h}\,(\epsilon(t), t)$$

are used to propagate the state forward in time, and to calculate the predicted observations from that state.

The extended Kalman filter equations are presented without proof using the following notation: suffix $(i|j)$ indicates that the variable is evaluated at time t_i using observations up to time t_j. Let n be the number of elements in the state vector, and m the number of elements in the observation vector. Table 9.9 defines the variables used in the filter equations.

Table 9.9

Extended Kalman Filter Variables

Variable	Description	Dimension
$\epsilon\,(k\|j)$	State vector estimate	$n \times 1$
$\mathbf{P}\,(k\|j)$	State error covariance matrix	$n \times n$
$\phi\,(k, j)$	State transition matrix from t_j to t_k	$n \times n$
$\mathbf{Q}\,(k)$	State noise covariance matrix at t_k	$n \times n$
η_k	Measurement vector at t_k	$m \times 1$
$\mathbf{H}\,(k)$	Measurement matrix at t_k	$m \times n$
$\mathbf{R}\,(k)$	Measurement noise covariance matrix at t_k	$m \times m$
$\mathbf{K}\,(k)$	Kalman gain matrix at t_k	$n \times m$

The procedure consists of two steps. In the first, the state vector, and state error covariance matrix are evaluated for time t_{k+1} using observations up to time t_k.

$$\epsilon\,(k+1|k) = \epsilon\,(k|k) + \int_{t_k}^{t_{k+1}} \mathbf{f}\,(\epsilon\,(t|t_k), t)\, dt \tag{8}$$

$$\mathbf{P}\,(k+1|k) = \phi\,(k+1, k)\,\mathbf{P}\,(k|k)\,\phi^{\mathrm{T}}(k+1, k) + \mathbf{Q}\,(k+1)$$

The second step improves the estimate of the state vector by using the observation at time t_{k+1}, and adjusts the state error covariance matrix accordingly. The equations are:

$$\mathbf{K}\,(k+1) = \mathbf{P}\,(k+1|k)\,\mathbf{H}^{\mathrm{T}}\,[\mathbf{H}\,\mathbf{P}\,(k+1|k)\,\mathbf{H}^{\mathrm{T}} + \mathbf{R}\,(k+1)]^{-1}$$

$$\epsilon\,(k+1|k+1) = \epsilon\,(k+1|k) + \mathbf{K}\,[\eta_{k+1} - \mathbf{h}\,(\epsilon\,(k+1|k), t_{k+1})]$$

$$\mathbf{P}\,(k+1|k+1) = [\mathbf{I} - \mathbf{K}\,\mathbf{H}]\,\mathbf{P}\,(k+1|k)\,[\mathbf{I} - \mathbf{K}\,\mathbf{H}]^{\mathrm{T}} + \mathbf{K}\,\mathbf{R}\,(k+1)\,\mathbf{K}^{\mathrm{T}}$$

where $\mathbf{H} \equiv \mathbf{H}\,(k+1)$, and $\mathbf{K} \equiv \mathbf{K}\,(k+1)$.

The procedure must be initialised with an *a priori* estimate ϵ_0, and its error covariance $\mathbf{P}_0$, where $\mathbf{P}_0 = E\left\{\epsilon_0 \epsilon_0^{\mathrm{T}}\right\}$. The state, and observation, noise covariance matrices, **Q** and **R**, must also be supplied. Note that the term:

$$\eta_{k+1} - \mathbf{h}\,(\epsilon(k+1\,|k),\, t_{k+1}) \qquad (9)$$

is the observation residual, $\mathbf{r}\,(k+1\,|k)$, at time t_{k+1}.

Owing to errors associated with the model of the dynamical system, its linearisation, its discretisation, and its computation, the state vector estimate produced by the filter may progressively diverge from the true value. This behaviour is called *filter divergence*. It arises when the state vector error covariance matrix produced by the filter is no longer representative of the true error covariance matrix. A characteristic of divergence is that the computed error covariance matrix becomes too small, and so the filter *thinks* it has calculated the state vector much more accurately than it actually has. The filter gain matrix, **K**, becomes very small, and new observations no longer serve to improve the estimate of the state vector. The state vector estimate, propagated by a system model containing errors, will progressively diverge from the true value without the corrective influence of recent observations.

The behaviour of the filter can be monitored by examining the observation residuals expressed by Eqn. (9). When the filter is behaving correctly, the residuals have zero mean value, i.e.:

$$E\left\{\mathbf{r}(k+1\,|k)\right\} = 0$$

When the filter begins to diverge, the residuals become inconsistent with the observation covariance matrix $\mathbf{R}(k+1)$, and they begin to show a non-zero mean value.

It is necessary, therefore, to introduce some means of preventing the state error covariance matrix, **P**, from decreasing too rapidly. A simple means of doing this is through the state noise covariance matrix, **Q**, which must be supplied to the filter. It is, unfortunately, very difficult to get a good estimate of **Q**. Its value is usually found by trial-and-error methods, using simulations of the orbit determination procedure. This method of compensating for the modelling error has been shown to be successful in suppressing filter divergence.

Let us now consider the linearisation of the state model. The dynamical equations are non-linear, and must include the influence of perturbations to simple two-body motion, if the physical system is to be modelled accurately. When the state is being propagated forwards (as in Eqn. (8)), the function $\mathbf{f}\,(\epsilon(t|t_k),\, t)$ must include the effect of the perturbations if good accuracy is to be achieved. In the calculation of **F** (in Eqn. (3a)), however, the *linearised* dynamical equations are used. It is valid, therefore, to exclude the perturbation terms from the dynamical equations used to calculate **F**. This considerably simplifies the calculation of the Jacobian matrix, **F**, and the state transition matrix, ϕ.

Let us finally consider the observation model. The observations are characterised by the observation noise covariance matrix **R**, corresponding to Gaussian white noise. No systematic, or bias, errors are modelled. If these are actually inherent in the observations, they must be estimated and removed before the observations can be used within the filter procedure. Fortunately, this is usually possible through a combination of analysis and testing.

9.3.3 Orbit Determination Error Estimation

During the mission design for a spacecraft, it is important to assess the orbit determination accuracy achievable by a tracking system, to ensure that it meets the requirements of the mission. In this section we shall discuss how this is carried out.

The technique used is very similar to that used for orbit refinement, in that the method of least squares is used.

Let η_k be the error in observation k and let the total number of observations be K. Let the station location error vector be $\mathbf{\Delta L}$, and let instrumentation and environmental errors contribute the vector $\mathbf{\Delta I}$ to the measurement error matrix $\mathbf{\Delta Y}_{(K\times 1)}$. We have, therefore:

$$\mathbf{\Delta Y}_{(K\times 1)} = \mathbf{A}_{(K\times N)} \cdot \Delta\epsilon_{(N\times 1)} + \mathbf{B}_{(K\times 3)} \mathbf{\Delta L}_{(3\times 1)} + \mathbf{C}_{(K\times Q)} \mathbf{\Delta I}_{(Q\times 1)} \quad (1)$$

where:

$\Delta\epsilon_{(N\times 1)}$ is the vector expressing the errors in the orbit parameters;

$\mathbf{A}_{(K\times N)}$ is the matrix of partial differentials as in Eqn. (1) of Section 9.3.2;

$\mathbf{B}_{(K\times 3)}$ is the matrix relating station location errors to errors in the measurements;

$\mathbf{C}_{(K\times Q)}$ is the matrix relating the instrumentation and environmental errors to the errors in the measurements. Note that Q is the number of error sources included.

Consider first noise errors only. Then Eqn. (1) above can be simplified to:

$$\mathbf{\Delta Y}_{(K\times 1)} = \mathbf{A}_{(K\times N)} \Delta\epsilon_{(N\times 1)} + \Omega$$

where Ω is the error contribution due to mismodelling and linearisation.

We now follow a procedure identical to the least squares approach of Section 9.3.2.1 Eqns. (2) through to (3), and we obtain the relation:

$$\Delta\epsilon = (\mathbf{A}^{\mathrm{T}} \mathbf{W}^{-1} \mathbf{A})^{-1} (\mathbf{A}^{\mathrm{T}} \mathbf{W}^{-1} \mathbf{\Delta Y}) \quad (2)$$

Now let $\mathbf{N}$ be a matrix representing the random noise in the observations. Then, for the unbiased estimate being considered here:

$$\Delta Y = N \quad ,$$

and the expected value of $\Delta Y \Delta Y^T$ is given by:

$$E(\Delta Y \Delta Y^T) = E(N N^T) = W$$

where **W** is the diagonal matrix of the variances of the observations, and is identical to the matrix **W** defined in Section 9.3.2.1.

If we now consider bias errors in addition to noise errors, we can rewrite Eqn. (2) in the form:

$$\Delta \epsilon = (A^T W^{-1} A)^{-1} [A^T W^{-1} (N + B \Delta L + C \Delta I)] \tag{3}$$

Now, since **N**, **ΔL**, and **ΔI** are statistically independent, as they arise from unrelated sources, we can write:

$$E(N \Delta L^T) = E(N \Delta I^T) = E(\Delta L \Delta I^T) = 0$$

Evaluating the expected value of $\Delta\epsilon\Delta\epsilon^T$ and using Eqn. (3) we obtain:

$$E(\Delta\epsilon\Delta\epsilon^T) = (A^T W^{-1} A)^{-1} [A^T W^{-1} (W + B\, E(\Delta L \Delta L^T)\, B^T + C\, E(\Delta I \Delta I^T)\, C^T)\, W^{-1} A] (A^T W^{-1} A)^{-1} .$$

Setting $E(\Delta L \Delta L^T) = W_L$ and $E(\Delta I \Delta I^T) = W_I$

We have:

$$E(\Delta\epsilon\Delta\epsilon^T) = (A^T W^{-1} A)^{-1} [A^T W^{-1} (W + BW_L B^T + CW_I C^T) W^{-1} A] (A^T W^{-1} A)^{-1}$$

Now $E(\Delta\epsilon\Delta\epsilon^T)$ is the covariance matrix of errors arising in the orbit parameter set ϵ, owing to the inaccuracies in the tracking system. Knowledge of the covariance matrix indicates the adequacy of the tracking system for the requirements of the mission.

9.4 Further Reading

Excellent treatments on orbit determination and tracking are found in the following documents, of which the first is also a very good primer on astrodynamics – the short historical introductions to the various chapters make most entertaining reading.

(a) Bate, R.R., Mueller, D.B., and White, J.E. : 1971, "Fundamentals of Astrodynamics", Published by Dover Publications Inc., New York.

(b) Soop, E.M. : 1983, "Introduction to Geostationary Orbits", *ESA SP-1053*, November 1983.

(c) Escobal, P.R. : 1965, "Methods of Orbit Determination", Published by John Wiley & Sons Inc., New York.

Orbit estimation using the Kalman filter technique is very well covered in:

(d) Chodas, P. : 1981, "Application of the Extended Kalman Filter to Several Formulations of Orbit Determination", *UTIAS Technical Note* No.224.

(e) Raol, J.R. and Sinha, N.K. : 1985, "On the Orbit Determination Problem", *IEEE Transactions on Aerospace and Electronic Systems*, Vol.AES-21, No.3, May 1985.

Tracking error estimation is covered in the following papers:

(f) Walter, H.G. : 1971, "Accuracy Estimation for Spacecraft Position and Velocity", *ESRO SR-16 (ESOC).*

(g) Vonbun, F.O., and Kahn, W.D. : 1962, "Tracking Systems, Their Mathematical Models and Errors. Part 1: Theory", *NASA TN D-1471.*

(h) Vonbun, F.O., and Kahn, W.D. : 1966, "Tracking Systems, Their Mathematical Models and Errors. Part 2 : Least Squares Treatment", *NASA TN D-3776.*

(i) Kawase, S., and Soop, E.M. : 1986, "Ground Antenna Pointing Performance for Geostationary Orbit Determination", *ESA Journal*, Vol.10, pp.71–83.

CHAPTER 10

SPACECRAFT STABILITY

10.1 Introduction

With the increasing use of liquid bi-propellant systems for spacecraft propulsion, consideration of spacecraft stability has become more important, and considerably more difficult. In this chapter we shall give an overview of the problems associated with the stabilisation of spacecraft containing large amounts of liquid propellants. The subject is a difficult one, and currently not all the problems are amenable to analytic investigation. To obtain confidence in the design, testing is necessary. We shall outline some of the test techniques available.

Before proceeding with the discussion of spacecraft stability during the various phases of the mission, we shall explain some of the terms that will be used.

Inertia ratio : This is the ratio of the moments of inertia about the spin axis and a transverse axis the origin of the axis system being at the spacecraft's centre of gravity.

Nutation angle : The instantaneous angle between the spacecraft's spin axis and the direction of the angular momentum vector.

Oblate : The spacecraft is said to be oblate if it spins about its maximum inertia axis.

Prolate : The spacecraft is said to be prolate if it spins about its minimum inertia axis.

Flat Spin : A condition when the spacecraft spins about an axis perpendicular to the desired spin axis.

Time Constant : The time it takes for the nutation angle to change by a factor of 2.718282.

10.2 Stability in Parking Orbit

Traditionally, spacecraft employing solid propellant apogee motors have adopted spin stabilisation during the parking orbit and transfer orbit phases of the mission. In transfer orbit, the spacecraft needs only to be spinning about its maximum inertia axis for passive stability to be assured. The spin speed is then selected to ensure that with possible thrust vector and centre of gravity misalignments, the nutation angle during the apogee boost phase remains within acceptable limits. This ensures that the apogee manoeuvre is executed with acceptable accuracy.

For *Shuttle* launch, however, during the duration of the parking orbit – typically 45 minutes – the spacecraft is attached to the upper stage which will inject it into geostationary transfer orbit. This almost certainly means that the spacecraft-plus-upper stage stack will be spinning about a minimum inertia axis. This is an unstable configuration, and if uncontrolled, will eventually lead to the stack spinning about its maximum inertia axis, though a progressive increase in the nutation angle. See Figure 10.1. This is called a *flat spin*, and, for a spacecraft, effectively means loss of the mission, because the perigee motor will not be pointing in such a way as to enable it to deliver the proper propulsive impulse to the spacecraft. To avoid this situation, active nutation control is instituted. This effectively counteracts the natural growth of the nutation angle and maintains the stack spinning about the minimum inertia axis. For spacecraft with only a small amount of liquid propellant on board, the rate of increase of the nutation angle in the absence of control, is quite low. This means that the nutation controller need only possess moderate authority to control the nutation angle effectively. More importantly, the nutation controller need only have a small capture envelope for it to function effectively. The significance of this needs explanation, and is related to the *Shuttle* payload deployment sequence.

To ensure the safety of the Orbiter and its crew, deployed payloads are prohibited from using their thrusters until they are outside a safety boundary around the Orbiter. The distance of this boundary around the Orbiter depends upon the nature of the thruster, and can shrink to zero for small cold-gas thrusters. For hot-gas thrusters commonly used on spacecraft, however, this safety boundary is about 200 ft. from the Orbiter, though this distance is negotiated by NASA on a case-by-case basis.

Most spin-stabilised payloads are spring-ejected from the cargo bay, and therefore have a finite separation velocity. This is of the order of 1 to 2 ft/s. It can be seen, therefore, that there exists a period of about 100s after deployment from the cargo bay during which thruster operation is forbidden. If the spacecraft's active nutation controller is based on thruster actuation (as most are), this period is effectively *dead time* for such a system. The nutation of the stack will, therefore, have grown to some finite value before the controller can begin to counteract it. The uncontrolled growth of the nutation angle depends upon the initial nutation angle and the *time constant* of nutation angle divergence. The former depends upon stack inertia characteristics and deployment system asymmetries, the latter depends largely upon the rate of energy dissipation within the stack. The greater the energy dissipation, the lower the time constant, and the faster the nutation angle increases.

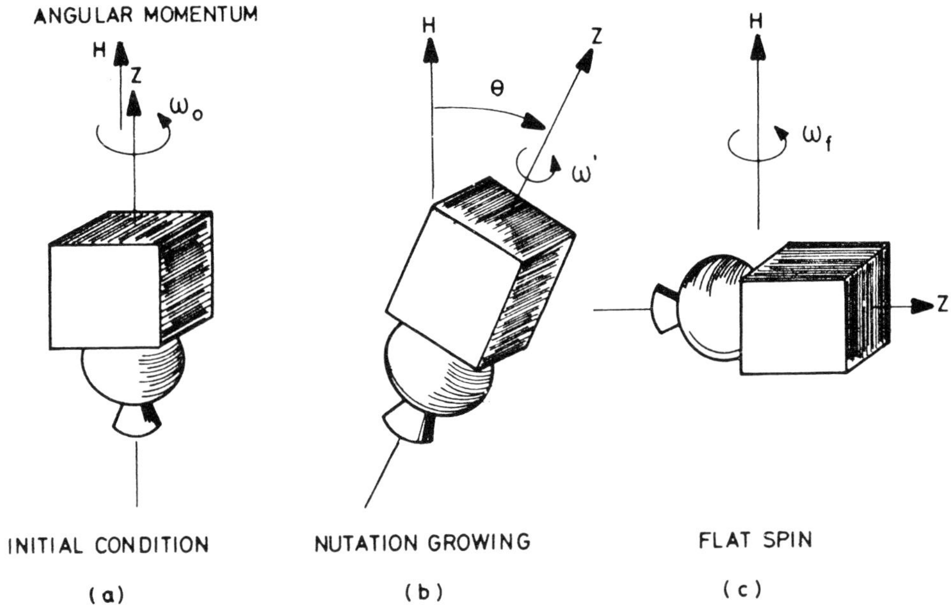

Fig.10.1 The Flat Spin

The energy dissipation of the stack comes from two main sources, fluid motion within the spacecraft's tanks, and structural movement due to the flexibility of certain elements of the spacecraft. For most current spacecraft and those of the near future, energy dissipation due to liquid motion is the more important effect of the two.

It can be seen, therefore, that when the liquid mass is low, the time constant will generally be longer, and the nutation angle generally smaller at the start of active nutation control, than when the liquid mass is high. Active nutation control is, therefore, generally easier to implement when the quantity of fluid on board the spacecraft is small. Spacecraft which use liquid propulsion for the apogee manoeuvre have, typically, about six times as much propellant as spacecraft which use solid propulsion for the apogee manoeuvre. The task of active nutation control becomes much more difficult for such spacecraft, and the active nutation controller must exert a much higher level of authority.

Design of the active nutation controller requires a good estimate of the *time constant* of nutation angle divergence, and it is in the prediction of this parameter that significant difficulty arises. Two approaches are available to the designer, i.e. analysis and test. In practice, shortcomings in both areas require a synthesis of the two, with greater reliance being placed upon test results.

10.3 Analytic Methods

The analytic methods currently in use are based upon the techniques outlined below:

- mechanical analogues simulating observed behaviour;
- fluid behaviour analysis based upon finite elements;
- fluid behaviour analysis based upon finite differences.

Much work has been done in all these areas, but a method with adequate predictive power still proves elusive.

10.4 Test Methods

The alternative is testing. Testing is complicated by the requirement to replicate zero-gravity effects on the ground. For test results to be meaningful, the liquid free surface within the tanks in the zero-gravity environment must be faithfully reproduced. Furthermore, the test article must be given six degrees of freedom in movement for the nutational and centre-of-gravity motion to be allowed to develop. Furthermore, the considerable difficulties inherent in full-scale testing sometimes make it necessary to work on reduced scale models. This introduces complexities of its own in the correct scaling of the model. Not only must the geometries and inertias be correctly scaled, but the parameters of importance in the motion of the liquid must also be scaled. Dimensional analysis indicates which parameters need to be scaled, but satisfying all the scaling requirements and respecting practical constraints sometimes produces conflict that must be carefully resolved. Producing the model for a nutation test is a demanding and skilled undertaking.

Several test techniques are available. These are:

- air bearing test,
- drop test,
- test within an aircraft flying a zero-gravity trajectory,
- spin-table/gimbal tests.

Of the above, the most used for spacecraft testing are the air-bearing test and the drop test.

The Air Bearing Test Technique

For this type of test, a suitably scaled model with liquid-containing tanks is set up on an air bearing, with the air bearing at the centre of gravity position of the model. The assembly is spun up, the air bearing activated, and the model released.

The nutation grows until it reaches the maximum value permitted by the rig, and the model is then captured. Accelerometers mounted on the model sense the transverse angular rates and provide the information from which the time constant can be deduced. A schematic representation of an air bearing rig is shown in Figure 10.2.

To avoid the effect of aerodynamic damping, the rig is usually mounted in a vacuum chamber. This both increases the cost of the testing, and slows it down considerably, as the chamber must be re-pressurised every time the model needs attention (for parameter changes), and then evacuated again for the next test run. Another drawback with this method is that it is not possible to test a model with tanks on the spin axis, because generally with such configurations the model centre of gravity position is occupied by a tank, or is otherwise inaccessible to the air bearing.

Where a model is feasible, the method is known to give consistent results, and comparison with on-orbit behaviour is reasonably good. The method has been used to good effect by a number of spacecraft manufacturers, and had the advantage that for the previous generation of spacecraft, at any rate, full scale testing was possible. This neatly side-stepped the problem of accurate scaling of the propellant expulsion device within the tank. Full scale *development model* tanks were generally used on the models. For modern spacecraft with very large propellant quantities on board, this advantage has disappeared, and scale models must be used.

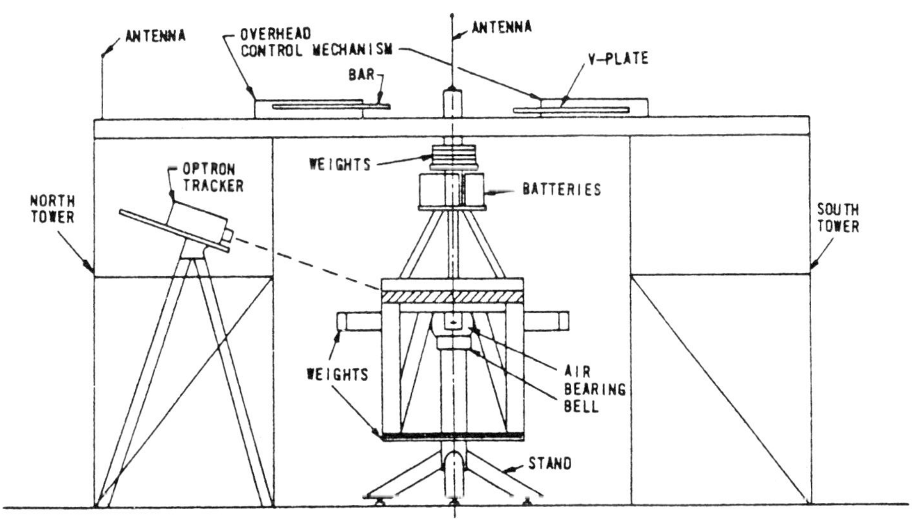

Fig.10.2 An Air Bearing Rig

The Drop Test Technique

The drop test technique has become increasingly popular for the new generation of spacecraft. Here a scale model (generally a geometric scaling of around 5:1 is used) is suspended in a rig with the spin axis horizontal. It is then spun up to a scaled spin rate and released from the rig. A schematic drawing of the spin-up and release mechanism is shown in Figure 10.3. As the model falls it nutates, and accelerometers within the model telemeter the transverse angular rates to receiving equipment on the ground. At the end of the drop, the model is caught in a box filled with foam rubber. The test is non-destructive, and drop heights of about 20 ft. are used, giving about a second of free-fall time. A typical drop tower is illustrated in Figure 10.4.

For the drop test results to be properly representative of the flight situation, the model must be carefully scaled with respect to several dynamical and fluid motion parameters. Table 10.1 shows which parameters need to be scaled and which parameters need to be identical, between the spacecraft and the test model. The first step in the scaling procedure is the choice of the liquids to represent the spacecraft's propellants. The liquids must reflect, as closely as possible, the density and kinematic viscosity of the propellants. It has been found that a water/methanol mixture is a suitable simulant for monomethylhydrazine, and that dichloromethane is a suitable simulant for nitrogen tetroxide. The relevant properties of the propellants and their simulants are shown in Table 10.2.

Table 10.1

Scaling Parameters

Parameter	Scaled	Identical
Inertia about principal axes	√	
Inertia ratio		√
Mass	√	
Dimensions of tank, tank location, PMD, model c.g	√	
Liquids : density, kinematic viscosity	√	
Spin rate, time	√	
Ratio of liquid-to-system mass		√
Ratio of liquid-to-system spin inertia		√
Tank fill fraction		√
Reynolds Number		√
Froude Number	greater than about 40	
PMD deflection due to liquid motion		√
PMD deflection due to spin field		√

Note: PMD is Propellant Management Device

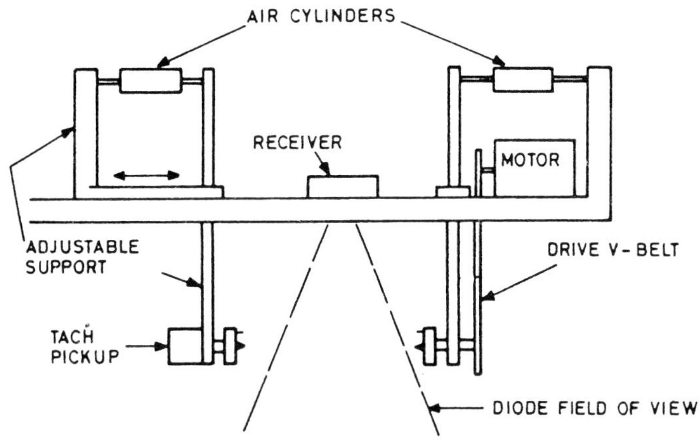

Fig.10.3 Spin-Up and Release Mechanism

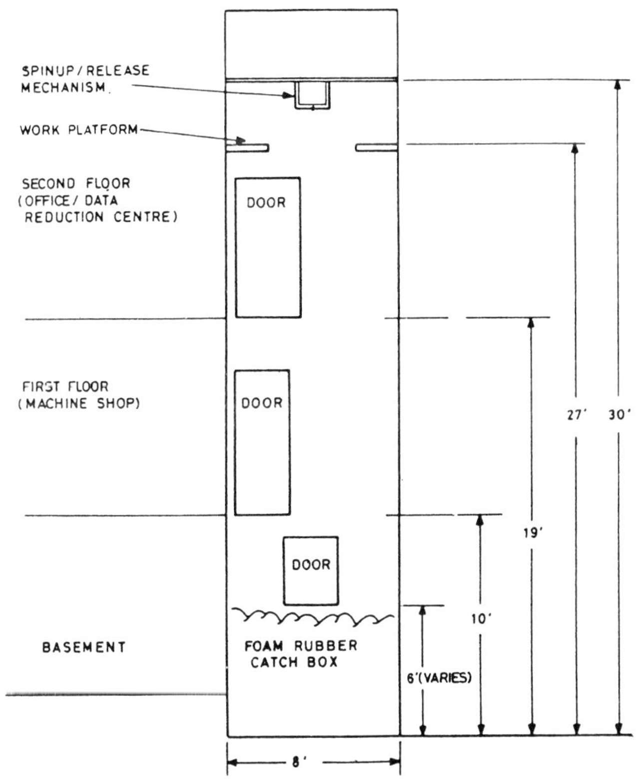

Fig.10.4 Drop Tower

Table 10.2

Propellant and Simulant Characteristics

	MMH	Water/Methyl Alcohol 94% Methanol 6% Water	N_2O_4	Dichloromethane
Density (kg/m^3)	878	811	1440	1330
Kinematic Viscosity (m^2/s)	1.02×10^{-6}	0.927×10^{-6}	0.28×10^{-6}	0.323×10^{-6}

It has been found that, provided the Reynolds Numbers of the spacecraft and the model are on the same side of a critical value, the test results are not sensitive to small differences in the Reynolds Number, i.e. of the order of 10%. The critical Reynolds number is believed to be that which determines whether the boundary layer of the fluid on the tank walls is laminar or turbulent. It is dependent upon the surface roughness characteristics of the tank walls, and can be expected to be in the region of 0.5×10^6 for smooth-walled tanks.

The Froude number is also a parameter of interest, and expresses the ratio between centrifugal acceleration and gravitational acceleration. For an orbiting spacecraft, its value is infinity. For the model, Froude numbers of about 40 or greater are perfectly adequate to ensure results representative of the spacecraft.

Finally, the scaling is such that the time constant of the model is very much smaller than that of the spacecraft. This is a valuable attribute, and allows observation of a sufficient number of nutation cycles within the available drop time. For typical model sizes, the time scaling between the spacecraft and the model is in the region of 50:1.

The general accuracy and repeatibility of the method are good. Comparison with flight tests indicates that the results are accurate to within ±15%. The method's main attraction lies in its relative cost-effectiveness and in the rapidity with which runs can be carried out. Wide parameter variations can be explored for a minimum expenditure of cost and time. Its only significant drawback is in the testing of spacecraft whose time constants are expected to be very long. The limited drop time would then result in too few nutation cycles being recorded, to give acceptable accuracy.

Tests in Aircraft Flying Zero-Gravity Trajectories

Aircraft have been used extensively to simulate zero-gravity conditions. For spacecraft nutation testing, the major drawbacks are the expense involved in the flights, and the safety considerations associated with testing a free-flying, spinning, fluid-filled model within the confines of an aircraft cabin. The technique has, however, been used successfully. With the alternative techniques available today this method can no longer be considered viable in terms of cost.

Spin Table Tests

Spin-table and gimball tests involve imposing the nutation motion of the spacecraft on to a fluid-filled container representing the propellant tank. The energy required to sustain the motion is monitored and recorded. This is seen as related to the energy dissipation of the fluid within the tank. Once the energy dissipation rate is determined, the *Energy Sink* theory is used to relate energy dissipation to the time constant of divergent (or convergent) nutation. The *Energy Sink* theory relates energy dissipation to the time constant through the formula:

$$\tau = \frac{\sigma(\sigma-1)\ I_T\,\omega^2}{(E/\theta^2)}$$

where τ = time constant,
σ = inertia ratio,
I_T = transverse inertia,
ω = spin rate,
E = instantaneous energy dissipation rate,
θ = instantaneous nutation angle.

The drawback of this method is that the accuracy of the derived time constant depends upon the validity of the *Energy Sink* relationship. Though this is expected to apply with good accuracy for small fluid masses, and small nutation angles, its validity is far from established when both are large. The method has been used more for fluid mechanics investigations that for spacecraft nutation testing, and gives generally consistent results.

10.5 Mission Design

Once the nutation time constant has been derived for all the relevant mission phases, the active nutation controller can be designed, and the mission sequence tailored to ensure the integrity of the spacecraft and the mission. This involves selection of the timing of sequencer operations, to respect the requirements of the spacecraft as well as to respect *Shuttle* safety requirements, and upper stage constraints. An example of such mission design is given below. The case considered is that of the EUROSTAR platform launched by *Shuttle*. The upper stage used for injection into transfer orbit is the McDonnell Douglas *PAM-D2*.

The EUROSTAR platform employs bi-propellant liquid propulsion for all necessary manoeuvres including the apogee manoeuvre. For this purpose it carries almost 1000 kg of propellant in four tanks symmetrically off-set from the spin axis, as shown in Figure 10.5. Extensive drop testing has shown that the spacecraft-plus-upper stage stack, in parking orbit, can be expected to exhibit a divergent nutation time constant of 174s at a spin rate of 40 rpm. To allow for errors in testing, this value has been factored down by 15% to give a time constant of 148s. This is a conservative approach, entirely suitable for preliminary mission design. The separation characteristics of the stack from the Orbiter's cargo bay indicate that the separation event could induce an initial nutation angle of 0.70° due to spring mismatch, Orbiter transverse rates, axis misalignment, etc. The separation springs give a separation velocity of about 2 ft/s, and hence the 200 ft. safety distance is reached in 100s.

If the active nutation damper were activated at this instant, it would at worst, have to cope with a nutation angle of 1.38°. However, the upper stage sequencer that is used to initiate the spacecraft-borne active nutation controller can only command the spacecraft at a finite number of pre-set times after separation. This enables the active nutation controller to become effective 140s after separation from the Orbiter. At this time it must cope with a nutation angle of 1.8°. In fact, to allow for system failure, detection of failure, and automatic switchover to a redundant system, the active nutation controller is designed to cope with a much greater nutation angle. In the case of EUROSTAR, the active nutation controller can cope with a nutation angle of about 7°. This allows ample safety margins and a robust and reliable mission. The EUROSTAR mission sequence at deployment from the Orbiter cargo bay is shown in Table 10.3.

10.6 Stability in Transfer Orbit

We now consider the situation that pertains at burn-out of the upper stage solid motor. At this time the spacecraft-plus-spent upper stage stack is in transfer orbit. During the burn of the upper stage motor, thrust vector misalignments and centre of gravity location errors cause an increase in the nutation angle of the stack. The nutation angle continues to grow up to the instant of separation of the spacecraft from the spent stage. If the angle is large, then the risk of contact between the spacecraft and the separated upper stage is high. Even if safe separation were obtained, the spacecraft, after separation, would suffer a pointing error of the spin axis which may violate power subsystem and thermal control constraints. It is important, therefore, to predict what the worst case nutation angle at separation will be. To do this, three pieces of information are required: (1) the nutation angle at burn-out, (2) the time constant of divergent nutation, and (3) the time between burn-out and separation.

If the nutation angle at separation is large, and its consequences regarding separation safety and pointing error are unacceptable, the mission designer has to consider other separation methods. One such method is to separate *after* the stack is established in a flat spin. Another is to initiate active nutation control after burn-out to reduce the nutation angle to an acceptably low value before separation.

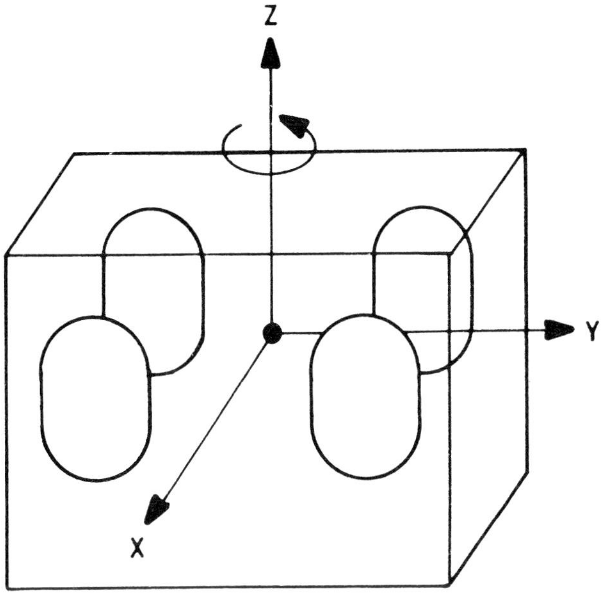

Fig.10.5 Tank Configuration

Table 10.3

EUROSTAR Deployment Sequence

Time (min:sec)	Event
T-60	Turn on PAM-D2 power
T-40	Start Orbiter manoeuvre to deployment attitude
T-28	Achieve deployment attitude
T-15	Start mechanical sequence
T-14:28	Restraint arms withdrawal complete
T-11:47	Spin-up complete (spin speed = 40 rpm)
T-11	Spacecraft ASE on
T-9	Spacecraft configured
T-7	Spacecraft to internal power
T-3	Start terminal sequence
T-0:00	Deploy PAM/spacecraft: sequencer initiated. separation velocity = 2 ft/s
T+0:100	Safe separation distance (200 ft.) reached, initialise CPS for start of AND
T+0:140	AND starts to provide control
T+44:58	Switch off AND
T+45:00	Fire PAM motor
T+46:49	PAM motor burn-out (burn time ~ 109s)
T+48:20	Initiate PAM/spacecraft separation by firing separation bolt cutters
T+48:23	Fire tumble system yo-weight bolt cutters.

Let us consider the first option. If, after burn-out, the stack remains uncontrolled, the nutation angle will increase until it reaches a value of 90°. At this point the stack will be in a flat spin rotating about its axis of maximum inertia. This is a stable dynamical configuration, and a favourable one in which to initiate separation of the spent upper stage from the spacecraft. After separation, if the spacecraft is oblate about its intended spin axis (i.e. the desired spin axis is also the axis of maximum inertia), the nutation angle will increase once more until it reaches 90° again, at which point the spacecraft will be spinning about the desired spin axis with zero nutation about that axis. This is equivalent to the spacecraft, after separation, entering another flat spin, at the end of which it is spinning about the desired spin axis. It is to be noted that this behaviour is entirely passive, and requires no intervention either to initiate or terminate it. The only requirement is that the spacecraft be adequately oblate (even in the face of liquid re-locations) about the desired spin axis. As far as the mission designer is concerned, he needs to know at what time after burn-out the flat spin situation will be established, so that he can time the separation command appropriately if this is required to be automatic. He can, alternatively, uplink the separation command from the ground, having confirmed, through telemetry reception at a ground station, that the flat spin has been established. As the spacecraft is near perigee, however, ground contact could prove awkward to establish.

If, after separation, the spacecraft is required to be three-axis stabilised, it must first be de-spun with thruster actuations, and then erected in the appropriate attitude through successive Sun- and Earth-lock manoeuvres.

Let us now consider the second option. Here the spacecraft active nutation controller is used to reduce the nutation angle to an acceptably low value for separation. As the stack dynamical characteristics are very different from what they were in parking orbit, the parameters governing the functioning of the active nutation controller must be changed from the values that were used in the parking orbit phase of the mission. It is very possible that the active nutation controller thruster configuration used in parking orbit, no longer gives adequate control authority. In such an event, a different thruster configuration will need to be employed. The initiation of the active nutation controller would need to be automatic, and a timer would have to be used. The mission designer would need to know the burn duration of the motor, and the time required for the controller to reduce the nutation angle to some low threshold value suitable for separation. It can be seen that much complexity is involved, and the solution is not a very attractive one.

Let us now consider the stability of the spacecraft during the rest of the transfer orbit phase of the mission. This phase lasts for a few days, typically five, and the major event is the firing of the apogee engine to circularise the orbit. Two main stability modes are suitable, the spin-stable mode and the three-axis stable mode. In the three axis stable mode, the spacecraft's spin rate is zero, and stability is maintained by either thrusters, reaction wheels, or momentum wheels, or a combination of these. To maintain stability during the apogee manoeuvre, thrusters are used to counteract disturbance torques arising from thrust vector misalignment with respect to the centre of gravity. If more than one apogee engine is used, thrust off-modulation, together with thruster control could provide attitude stability. For spacecraft, thrust vector gimballing it not favoured, mainly due to the mass penalty involved in such a system. Gyros are used to sense the attitude drift, and to provide control of the attitude control actuators. Earth and Sun sensors provide the means of periodically updating the gyros and cancelling their drift.

Consider now the spin-stabilised mode. If such a mode is adopted, it is most efficient for the spin axis to be the axis of maximum inertia. This provides passive stability and is inherently safe and reliable. If the spin axis is not the axis of maximum inertia, active nutation control must be provided throughout this phase, and the capability must be provided to recover from a "flat spin" condition. Further, careful inertia control must be instituted in the design phase to ensure that the inertia ratios (ratio of inertia about the spin axis to the inertia about each of the transverse axes) do not get too close to unity. However, an advantage with such a configuration is that the spacecraft is stable with respect to wobble (coning of the spacecraft about the spin axis, owing to mass imbalance).

There is an important class of spacecraft called dual-spin spacecraft wherein one part of the spacecraft can be imparted a spin rate relative to the other. Even if the spin configuration of such a spacecraft is prolate (i.e. it is spinning about the minimum inertia axis), it can be rendered passively stable by "super-spin", i.e. increasing the spin rate of one part of the spacecraft with respect to, and at the expense of, the other.

Consider now a spacecraft with a large amount of liquid propellant on board which is stabilised in transfer orbit by spinning about the axis of maximum inertia. Such a spacecraft is prone to wobble instability unless the inertia ratios are sufficiently greater than unity. The instability is affected by the interconnection of propellant tanks such as to enable propellant from one tank to enter another. The instability is also affected by the ignition of the apogee engine as this causes relocation of the propellant within the tanks. Ensuring that the inertia ratios are sufficiently greater than unity, automatically ensures spin stability. The mission designer needs, however, to determine the minimum inertia ratio to ensure stability. It is not always possible to do this by testing, and analytic stability analyses are necessary. These are generally based on Liapunov's stability criterion. Once the minimum inertia ratio has been determined, the spacecraft designer must ensure that this limit is respected, otherwise active nutation control must be provided.

10.7 On-Station Stability

Now consider the spacecraft to be on station and operational in geostationary orbit. Let its attitude stabilisation mode be three-axis stabilised and Earth-pointing. Most of the liquid propellant has been expended in executing the apogee manoeuvre, and what remains suffices for attitude control, and station-keeping through the spacecraft's operational lifetime. During the periods when the spacecraft's thrusters are not firing, i.e. over 99% of the on-station mission duration, the propellant within the tanks is in an orientation determined by surface tension forces. When thrusters are actuated, the spacecraft is under acceleration, and the propellant relocates. This relocation leads to a movement of the spacecraft's centre of gravity, and results in the thrusters causing a perturbation in the spacecraft's attitude, which could violate the pointing accuracy requirements of the mission. To ensure that this does not happen, the control system must be designed to counteract the disturbing torques.

This requires them to be quantified in terms of amplitude, frequency and damping. Since the disturbance torques arise due to the effects of transient liquid motion within a container of complex geometry, it can be seen that the problem is a most demanding one. Since testing is not feasible, analytic methods are used. These are capable of assessing the problem, but must be used with care, and a conservative approach is to be recommended. This would involve making simplifying worst case assumptions and designing to these.

10.8 Conclusions

In conclusion, it can be seen that consideration of spacecraft stability is central to the integrity of the mission, and much effort needs to be invested in ensuring that adequate stability margins exist throughout the mission.

The existence of large quantities of fluid within the spacecraft has a major influence on stability, and must be fully taken into account to ensure a safe mission. The analytic tools available to do this are sometimes inadequate and adequate confidence in the design is only obtained through extensive testing. This testing should be carried out as early as possible during the spacecraft design phase in order to ensure that enough time is available to solve problems should these arise.

Some of the most difficult problems can entirely be avoided if the launch vehicle used is *Ariane*, and if the spacecraft configuration is adequately oblate. *Ariane* is three axis stabilised throughout its flight. This avoids the spin stability problems associated with a spinning upper stage. Once the spacecraft has separated from the launch vehicle, its passive spin stability may be assured by designing a sufficiently oblate configuration.

It is clear that by a combination of launch vehicle selection, spacecraft configuration design, and adequate testing and analysis, spacecraft stability can be assured throughout the mission.

10.9 Further Reading

References (a) to (d) outline theoretical attempts to model the nutation dynamics of bodies containing substantial quantities of fluids.

(a) Pfeiffer, F. : 1984, "On Contained Rotating Fluids in Satellite Dynamics", Proceedings of the Intelsat/ESA Colloquium on Dynamic Effects of Liquids on Spacecraft Attitude Control. Washington D.C.

(b) Pfeiffer, F. : 1977, "Problems of Contained Rotating Fluids with Respect to Aerospace Applications", Proceedings of the Conference on Attitude Control of Space Vehicles, *ESA SP-129*, pp.51–62.

(c) Guibert, J.P. : 1984, "An Approximate Solution for Propellant Slosh in a Spinning Tank", Proceedings of the Intelsat/ESA Colloquium on Dynamic Effects of Liquids on Spacecraft Attitude Control, Washington D.C.

(d) Pohl, A. : 1984, "Dynamic Effects of Liquids on Spinning Spacecraft", Proceedings of the Intelsat/ESA Colloquium on Dynamic Effects of Liquids on Spacecraft Attitude Control, Washington D.C.

References (e) to (g) describe experimental techniques for the characterisation of spacecraft nutation dynamics.

(e) Harrison, J.V. et al. : 1983, "A Free-Fall Technique to Measure Nutation Divergence, and Applications", AAS/AIAA Astrodynamics Specialist Conference, Lake Placid, New York, Paper No.83-372.

(f) Vanyo, J.P. : 1973, "An Energy Assessment for Liquids in a Filled Precessing Spherical Cavity", *Journal of Applied Mechanics*, Dec. 1973.

(g) Reiter, G.S. and Lee, D.A. : 1966, "Zero-Gravity Stability Testing of a Liquid Filled Space Vehicle", *Aerospace Chemical Engineering*, No.61, Vol.62, pp.178-183.

The final group of references discusses analytic methods for the evaluation of spacecraft stability under the influence of moving liquids.

(h) McIntyre, J.E. and Miyagi, M.I. : 1976, "A General Stability Principle for Spinning Flexible Bodies with Application to the Propellant Migration – Wobble Amplification Effect", Proceedings of the ESA symposium on the Dynamics and Control of Non-Rigid Spacecraft, Frascati, *ESA SP-117*, pp.159-176.

(i) Agrawal, B.N. : 1982, "Stability of Spinning Spacecraft with Partially Liquid-filled tanks", *J. of Guidance*, Vol.5, No.4, pp.344-350.

(j) Metzger, R. : 1977, "Stability Problems of a Liquid Propellant Apogee Motor", Proceedings of the Conference on Attitude Control of Space Vehicles, *ESA SP-129*, pp.143-156.

(k) Guibert, J.P. : 1983, "A Property of Modal Convergence for a Spinning Non-Rigid Body", *Acta Astronautica*, Vol.10, No.5, pp.347-355.

Appendix A

Including *zonal* harmonics from J_2 to J_7, and *tesseral* harmonics up to (S_{44}, C_{44}), the expressions for f_r, f_L and f_λ are as follows:

$$f_r = -\frac{\delta u'}{\delta r} = -\frac{\mu J_2 R^2}{r^4}\left(\frac{9}{2}\sin^2 L - \frac{3}{2}\right) - \frac{\mu J_3 R^3}{r^5}(10\sin^3 L - 6\sin L)$$

$$-\frac{\mu J_4 R^4}{r^6}\left(\frac{175}{8}\sin^4 L - \frac{150}{8}\sin^3 L + \frac{15}{8}\right) - \frac{\mu J_5 R^5}{r^7}\left(\frac{189}{4}\sin^5 L - \frac{210}{4}\sin^3 L\right.$$

$$\left. + \frac{45}{4}\right)\sin L$$

$$-\frac{\mu J_6 R^6}{r^8}\left(\frac{1617}{16}\sin^6 L - \frac{2205}{16}\sin^4 L + \frac{735}{16}\sin^2 L - \frac{35}{16}\right)$$

$$-\frac{\mu J_7 R^7}{r^9}\left(\frac{429}{2}\sin^7 L - \frac{693}{2}\sin^5 L + \frac{315}{2}\sin^3 L - \frac{35}{2}\sin L\right)$$

$$-\frac{\mu R^2}{r^4}(C_{22}\cos 2\lambda + S_{22}\sin 2\lambda)$$

$$-\frac{\mu R^3 \cos L}{r^5}\Big[(30\sin^2 L - 6)(C_{31}\cos\lambda + S_{31}\sin\lambda)$$

$$+ 60\sin L\cos L\,(C_{32}\cos 2\lambda + S_{32}\sin 2\lambda) + 60\cos^2 L\,(C_{33}\cos 3\lambda + S_{33}\sin\lambda)\Big]$$

$$-\frac{\mu R^4 \cos L}{r^6}\left[\left(\frac{175}{2}\sin^3 L - \frac{75}{2}\sin L\right)(C_{41}\cos\lambda + S_{41}\sin\lambda)\right.$$

$$+ \cos L\left(\frac{525}{2}\sin^2 L - \frac{75}{2}\right)(C_{42}\cos 2\lambda + S_{42}\sin 2\lambda)$$

$$+ 525\cos^2 L\sin L\,(C_{43}\cos 3\lambda + S_{43}\sin 3\lambda)$$

$$\left. + 525\cos^3 L\,(C_{44}\cos 4\lambda + S_{44}\sin 4\lambda)\right]$$

$$f_L = -\frac{1}{r}\frac{\delta u'}{\delta L} = \frac{\mu J_2 R^2}{r^4}(3 \sin L \cos L) + \frac{\mu J_3 R^3}{r^5}\left(\frac{15}{2}\sin^2 L \cos L - \frac{3}{2}\cos L\right)$$

$$+ \frac{\mu J_4 R^4}{r^6}\left(\frac{35}{2}\sin^3 L \cos L - \frac{15}{2}\sin L \cos L\right)$$

$$+ \frac{\mu J_5 R^5}{r^7}\left(\frac{315}{8}\sin^4 L \cos L - \frac{210}{8}\sin^2 L \cos L + \frac{15}{8}\cos L\right)$$

$$+ \frac{\mu J_6 R^6}{r^8}\left(\frac{693}{8}\sin^5 L \cos L - \frac{315}{4}\sin^3 L \cos L + \frac{105}{8}\sin L \cos L\right)$$

$$+ \frac{\mu J_7 R^7}{r^9}\left(\frac{3003}{16}\sin^6 L \cos L - \frac{3465}{16}\sin^4 L \cos L + \frac{945}{16}\sin^2 L \cos L - \frac{35}{16}\cos L\right)$$

$$+ \frac{\mu R^2}{r^4}(-6 \sin L - \cos L)(C_{22} \cos 2\lambda + S_{22} \sin 2\lambda)$$

$$+ \frac{\mu R^3}{r^5}\left[\left(\frac{45}{2}\cos^2 L - 6\right)\sin L\,(C_{31} \cos \lambda + S_{31} \sin \lambda) + \cos L\,(45 \cos^2 L - 30)\right.$$

$$\left.(C_{32} \cos 2\lambda + S_{32} \sin 2\lambda) - 45 \cos^2 L \sin L\,(C_{33} \cos 3\lambda + S_{33} \sin 3\lambda)\right]$$

$$+ \frac{\mu R^4}{r^6}\left[\left(-70 \sin^4 L + \frac{135}{2}\sin^2 L - \frac{15}{2}\right)(C_{41} \cos \lambda + S_{41} \sin \lambda)\right.$$

$$+ \cos L \sin L\,(210 \cos^2 L - 90)(C_{42} \cos 2\lambda + S_{42} \sin 2\lambda)$$

$$+ \cos^2 L\,(420 \cos^2 L - 315)(C_{43} \cos 3\lambda + S_{43} \sin 3\lambda)$$

$$\left. - 420 \cos^3 L \sin L\,(C_{44} \cos 4\lambda + S_{44} \sin 4\lambda)\right]$$

$$f_\lambda = -\frac{1}{r\cos L}\frac{\delta u'}{\delta\lambda} = \frac{\mu R^2}{r^4} 6\cos L\,(S_{22}\cos 2\lambda - C_{22}\sin 2\lambda)$$

$$+\frac{\mu R^3}{r^5}\Bigg[\left(\frac{15}{2}\sin^2 L - \frac{3}{2}\right)(S_{31}\cos\lambda - C_{31}\sin\lambda)$$

$$+ 30\sin L\cos L\,(S_{32}\cos 2\lambda - C_{32}\sin 2\lambda) + 45\cos^2 L\,(S_{33}\cos 3\lambda - C_{33}\sin 3\lambda)\Bigg]$$

$$+\frac{\mu R^4}{r^6}\Bigg[\left(\frac{35}{2}\sin^3 L - \frac{15}{2}\sin L\right)(S_{41}\cos\lambda - C_{41}\sin\lambda) + \cos L\,(105\sin^2 L - 15)$$

$$(S_{42}\cos 2\lambda - C_{42}\sin 2\lambda) + 315\cos^2 L\sin L\,(S_{43}\cos 3\lambda - C_{43}\sin 3\lambda)$$

$$+ 420\cos^3 L\,(S_{44}\cos 4\lambda - C_{44}\sin 4\lambda)\Bigg]$$

The values of the *zonal* and *tesseral* harmonics are:

$J_2 = -1082.64 \times 10^{-6}$
$J_3 = 2.541 \times 10^{-6}$
$J_4 = 1.618 \times 10^{-6}$
$J_5 = 0.228 \times 10^{-6}$
$J_6 = -0.552 \times 10^{-6}$
$J_7 = 0.352 \times 10^{-6}$

$C_{22} = 1.536 \times 10^{-6}$	$S_{22} = 0.882 \times 10^{-6}$
$C_{31} = 2.158 \times 10^{-6}$	$S_{31} = 0.241 \times 10^{-6}$
$C_{32} = 0.266 \times 10^{-6}$	$S_{32} = -0.258 \times 10^{-6}$
$C_{33} = 0.068 \times 10^{-6}$	$S_{33} = 0.213 \times 10^{-6}$
$C_{41} = -0.491 \times 10^{-6}$	$S_{41} = -0.457 \times 10^{-6}$
$C_{42} = 0.077 \times 10^{-6}$	$S_{42} = 0.150 \times 10^{-6}$
$C_{43} = 0.062 \times 10^{-6}$	$S_{43} = -0.007 \times 10^{-6}$
$C_{44} = 0.002 \times 10^{-6}$	$S_{44} = 0.008 \times 10^{-6}$

The confidence in the knowledge of these constants diminishes as the order of the constant increases.

INDEX

R

S